Wireless Networks

Series editor

Xuemin (Sherman) Shen
University of Waterloo, Waterloo, Ontario, Canada

More information about this series at http://www.springer.com/series/14180

Miaowen Wen · Xiang Cheng
Liuqing Yang

Index Modulation for 5G Wireless Communications

Springer

Miaowen Wen
South China University of Technology
Guangzhou, Guangdong
China

Xiang Cheng
School of Electronics Engineering
and Computer Science
Peking University
Beijing
China

Liuqing Yang
Electrical and Computer Engineering
Colorado State University
Fort Collins, CO
USA

ISSN 2366-1186 ISSN 2366-1445 (electronic)
Wireless Networks
ISBN 978-3-319-51354-6 ISBN 978-3-319-51355-3 (eBook)
DOI 10.1007/978-3-319-51355-3

Library of Congress Control Number: 2016960720

Printed on acid-free paper

This Springer imprint is published by Springer Nature
The registered company is Springer International Publishing AG
The registered company address is: Gewerbestrasse 11, 6330 Cham, Switzerland

Preface

The explosive increase of mobile data services and the use of smart phones require the fifth generation (5G) network to support higher spectral efficiency, higher energy efficiency, and higher mobility. The currently employed modulation techniques based on the multiple-input multiple-output orthogonal frequency division multiplexing (MIMO-OFDM) paradigm unfortunately fail to satisfy this requirement. The MIMO modulation achieves high spectral efficiency for a configuration of massive antennas, but brings about unsatisfactory energy efficiency due to scalable power consumption of radio frequency (RF) chains. The OFDM modulation is prone to intercarrier interference caused by Doppler effects and suffers from high peak-to-average power ratio (PAPR). New groundbreaking modulation technologies are therefore very demanding. The recently emerging index modulation techniques provide a breakthrough to the bottleneck. Index modulation refers to a family of modulation techniques that rely on the indices of some medium to convey additional information. The medium can be either practical such as antenna, subcarrier, and frequency carrier, or virtual such as time slot, space-time matrix, and antenna activation order. Unlike conventional modulation techniques, the information carried by index modulation is embedded into the actual transmitted signal, consuming little or even no power.

This monograph includes transceiver design, system optimization, and performance analysis of index modulation techniques in space, space-time, and frequency domains. The chapters of this monograph are relatively independent with each other. The readers can go directly to the chapter(s) of interest. This monograph starts with the motivation and a brief review of index modulation techniques in Chap. 1. The receiver-side spatial modulation and virtual spatial modulation techniques are introduced, followed by the applications to cooperative communications in Chap. 2. Both transmitter-side and receiver-side differential spatial modulation systems are studied, followed by the applications to cooperative communications in Chap. 3. Basic and enhanced index modulated OFDM systems are investigated, followed by the verification of their strong capabilities in high-mobility communications in Chap. 4. Finally, this monograph highlights some open problems and discusses future research directions in Chap. 5.

This monograph is designed for researchers in the field of wireless communications.

We would like to thank Ms. Yuyang Bian, Mr. Meng Zhang, Dr. Yuke Li, Dr. Jun Li, Dr. Beixiong Zheng, Dr. Zhiguo Ding, Dr. Ertugrul Basar, and Dr. Bingli Jiao for their much help in the research. We also want to express our thanks to Dr. Xilin Cheng for his valuable insights and comments. Finally, we would like to thank for the continued support from the National Natural Science Foundation of China under grants 61501190, 61622101, and 61571020, and the Natural Science Foundation under grant NSF-CNS 1343189.

Guangzhou, China Miaowen Wen
Beijing, China Xiang Cheng
Fort Collins, USA Liuqing Yang

Contents

Acronyms

ABEP	Average Bit Error Probability
AF	Amplify-and-Forward
ASDV	Adjacent Subcarrier Distance Vector
AWGN	Additive White Gaussian Noise
BER	Bit Error Rate
CD	Centralized Detection
CDF	Cumulative Distribution Function
CFO	Carrier Frequency Offset
CIOD	Coordinate Interleaved Orthogonal Design
CIR	Carrier-to-Interference Ratio
CP	Cyclic Prefix
CSI	Channel State Information
CSIR	Channel State Information at the Receiver
CSIT	Channel State Information at the Transmitter
DA	Differential Alamouti
DD	Distributed Detection
DH-HDSM	Dual-Hop Hybrid Differential Spatial Modulation
DH-HSM	Dual-Hop Hybrid Spatial Modulation
DN	Destination Node
DPSK	Differential Phase-Shift Keying
DSM	Differential Spatial Modulation
DSSK	Differential Space Shift Keying
DSTSK	Differential Space-Time Shift Keying
ESA	Equiprobable Subcarrier Activation
FBE	Fractional Bit Encoding
GPQSM	Generalized Precoding Aided Quadrature Spatial Modulation
GPSM	Generalized Precoding Aided Spatial Modulation
GSM	Generalized Spatial Modulation
ICI	Intercarrier Interference
IFFT	Inverse Fast Fourier Transform

LLR	Log-Likelihood Ratio
LTE	Long-Term Evolution
MIMO	Multiple-Input Multiple-Output
ML	Maximum-Likelihood
MMSE	Minimum Mean Squared Error
MSE	Minimized Mean Square Error
OFDM	Orthogonal Frequency Division Multiplexing
OFDM-HIM	OFDM with Hybrid Index Modulation
OFDM-IM	OFDM with Index Modulation
OFDM-SSK	OFDM With Subcarrier Shift Keying
OOK	On-Off Keying
PA	Power Allocation
PAPR	Peak-to-Average Power Ratio
PDF	Probability Density Function
PDSM	Precoding Aided Differential Spatial Modulation
PEP	Pairwise Error Probability
PSK	Phase-Shift Keying
PSM	Precoding Aided Spatial Modulation
QAM	Quadrature-Amplitude Modulation
R2V	Road to Vehicle
SAP	Subcarrier Activation Pattern
SM	Spatial Modulation
SN	Source Node
SSK	Space Shift Keying
SVD	Singular Value Decomposition
V2X	Vehicle to X
VSM	Virtual Spatial Modulation
ZF	Zero-Forcing

Chapter 1
Introduction

The fourth generation (4G) standardization and the ongoing worldwide deployment of 4G cellular network is maturing. However, with an explosion of wireless mobile devices and services, there are still some challenges that the 4G physical layer technologies cannot be accommodated. Technologists worldwide, thereby, have begun searching the 5G physical layer technologies to meet the anticipated demands in the 2020 era. The recently emerging index modulation appears as a promising candidate. In this chapter, first, we briefly address the 5G demand and identify key challenges. Then, we introduce the history of index modulation, mainly focusing on the space, space-time, and frequency domains realizations. Finally, the organization of this monograph is given at the end of this chapter.

1.1 5G Demands and Challenges

Wireless communications is the fastest growing segment of the communications industry. From around 1980 to the present, the wireless systems have traveled through 1G, 2G, and 3G. Nowadays, the long-term evolution (LTE) and its extension, LTE-advanced (LTE-A), as practical 4G systems, are steadily expanding across global markets. With the support of several key technologies, i.e., multiple-input multiple-output (MIMO) and orthogonal frequency division multiplexing (OFDM), cooperative communications, etc., 4G systems provide data rates of up to 1 Gb/s for low mobility and up to 100 Mb/s for high mobility. MIMO, which uses multiple antennas at the transmitter and receiver, boosts the channel capacity without satisfying any transmit power. Generally, there are three categories of MIMO techniques. The first one aims to maximize the spatial diversity by means of delay diversity, space-time block codes (STBC) [1], and space-time trellis codes (STTC) [2]. The second one uses a layered approach to transmit multiple independent data streams, such as vertical Bell Laboratories layered space-time (V-BLAST) [3]. Finally, the third one exploits the channel knowledge at the transmitter to perform beamforming. OFDM is a kind of multi-carrier modulation techniques, which converts a frequency-selective

M. Wen et al., *Index Modulation for 5G Wireless Communications*,
Wireless Networks, DOI 10.1007/978-3-319-51355-3_1

channel into a parallel collection of frequency flat subchannels. OFDM promises several advantages including the mitigation of inter-symbol interference, ease of implementation due to the adoption of fast Fourier transform (FFT), high spectral efficiency as the spectra corresponding to the orthogonal subcarriers overlap in frequency, and excellent pairing for MIMO. Cooperative communication techniques were first used in a distributed MIMO fashion and later to bridge connection with a destination node without a direct link, promising significant performance gains in terms of link reliability, spectral efficiency, system capacity, and transmission range.

The world, however, is also changing rapidly. With the development of industry and the worldwide deployment of mobile network, there is a dramatic increase in the number of users who subscribe to wireless systems every year. Meanwhile, powerful smartphones and laptops are becoming more and more popular. Driven by the video streaming and other various bandwidth-hungry applications, such as tele-presence and 3D holography, the amount of IP data handled by wireless networks is expected to have increased by over a factor of 100 to exceed 500 exabytes by 2020 [4]. On the other hand, as global CO_2 emissions increase and sea levels rise, global weather and air pollution are becoming more severe. Reports show that in 2012, telecom infrastructure and devices accounts for 25% of the annual average power consumption by information and communication technologies (ICT) industries [5]. All the above issues are already pushing the limits of current cellular systems. So far, it has been widely acknowledged that in comparison with the 4G wireless network, the 5G wireless network should achieve 1,000 times the system capacity, 10 times the spectral efficiency, energy efficiency, and data rate, and 25 times the average cell throughput [6]. In this context, revolutionary wireless communications technologies that are capable of catching up with the mobile data traffic growth while reducing the carbon emission are compelling.

To meet the requirement of high data rate, one may think of the appealing massive MIMO technique. With channel state information at the transmitter (CSIT), massive MIMO can be used to simultaneously serve a huge number of single-antenna users by sharing multiplexing gain among them. However, several problems, such as CSI overhead and requirement of multiple radio frequency (RF) chains, exist. Particularly, acquiring CSI in frequency division duplexing (FDD) systems is very costly due to immense information exchange. RF chains are in general very expensive and do not follow Moore's law [7]. More importantly, the associated circuits inside RF chains, which substantially consume power, significantly decrease the energy efficiency. Therefore, it is necessary to find other efficient schemes for (massive) MIMO that can work without CSIT and with less active RF chains while satisfying the high spectral efficiency demand. OFDM, as mentioned earlier, has impressive qualities, and thus is the unquestionable frontrunner for 5G. However, some weak points that could be more pronounced in 5G networks, such as high peak-to-average-power ratio (PAPR) and susceptibility to intercarrier interference (ICI), do exist. The PAPR problem arises from the aliasing of the frequency uncorrelated inputs, which becomes more severe as the number of FFT points increases. The ICI, which accompanies with the breaking of subcarrier orthogonality, goes stronger with Doppler effects. Therefore, it is of necessity to find other efficient schemes for OFDM that can solve these issues.

On the other hand, it is essential that the potential alternatives to MIMO and OFDM are applicable to cooperative communications, high-mobility communications, and other scenarios of interest.

1.2 Introduction of Index Modulation Family

Index modulation is an emerging signal modulation concept conceived for but not restricted to MIMO-OFDM communications. As its name suggests, index modulation modulates signals through the indices of some medium, which can be either practical, such as antenna, frequency carrier, and subcarrier, or virtual, such as time slot, space-time matrix, and antenna activation order. The indices are embedded into the transmitted or received signals, usually consuming little or even no power but carrying additional information. This implies that attractive tradeoff between spectral efficiency and energy efficiency can be achieved by index modulation. Moreover, inactivating some OFDM subcarriers for index modulation significantly relieves the PAPR and ICI problems. Those merits make index modulation a competitive candidate for 5G wireless communications.

1.2.1 Space Domain Index Modulation Overview

Spatial modulation (SM) is widely acknowledged to be the pioneer that realizes the concept of index modulation. SM belongs to the MIMO family, but it is actually an entirely new modulation technique. SM requires activation of only a single antenna at each time slot and exploits the uniqueness and randomness properties of the wireless channel for communications. The single-antenna-activation property allows SM to eliminate inter-channel interference and to avoid inter-antenna synchronization (IAS) while achieving a spatial multiplexing gain. To date, various works have demonstrated that SM has the potential to outperform some well-known MIMO techniques, such as the V-BLAST architecture [3], and amplitude/phase modulation with maximum ratio combining (MRC), in terms of error performance and energy efficiency, making it a promising candidate for 5G wireless communications [6, 7].

The first paper regarding SM appeared in 2002, where the proposed scheme was termed space shift keying (SSK) [8]. In this scheme, more than one antennas are switched on to transmit signals and the information is embedded into the differences in the form of signals. One year later, in [9] the idea of exploiting the position of the transmit antenna as information carrying unit was proposed. However, in this work only BPSK modulation was employed. An independent work of [9] with a similar idea was carried out in [10] after two years, where the proposed scheme was called channel-hopping modulation. This scheme was further completed with theoretical capacity analysis in 2008, and renamed as information-guided channel hopping (IGCH) modulation [11]. The year 2008 is crucial for the SM development since it is when the term SM was coined [12], an optimal maximum-likelihood (ML) detector was designed [13], and the advantages mentioned earlier were revealed [13].

In 2009, the concept of SSK modulation was redefined and clearly distinguished from SM [14]. Henceforth, SM indicates a modulation scheme in which transmitted symbols are drawn from both the spatial constellation and the signal constellation [15], while SSK modulation refers to only the spatial constellation. Therefore, SSK modulation can be considered as a special realization of SM. Despite a slight reduction of spectral efficiency, the elimination of the signal constellation in SSK modulation provides outstanding advantages over SM, e.g., simple transceiver framework, low detection complexity, and ease of integration within communication systems [14]. Later in 2010, the same authors of [14] proposed an improved version of SSK modulation, called generalized SSK (GSSK), by exploiting the antenna index combinations for communications [16]. Explicitly, GSSK improves the spectral efficiency at the price of the introduction of IAS. Similarly, by allowing activation of multiple antennas one can obtain the generalized SM (GSM). So far, two GSM versions are available in the literature: One transmits a single modulated symbol from multiple active antennas at each time slot, thus requiring only one RF chain [17], while the other one requires same number of RF chains as that of active antennas to transmit multiple independent modulated symbols [18].

To further explore the space resource for index modulation, a great deal of work following SM have been proposed. Quadrature SM (QSM) is one of the representative examples, which expands the spatial domain to include both the in-phase and quadrature dimensions [19]. In QSM, only a single RF chain is required to generate a complex modulated symbol, whose real and imaginary parts are transmitted by one/two active antennas chosen according to two independent bit streams, doubling the spectral efficiency contributed by the index bits. Another representative example is the so-called enhanced SM (ESM) [20]. In ESM, one or two antennas are activated for transmission at each time slot. In the case of a single active antenna, a primary signal constellation is used, while in the case of two active antennas, some other secondary constellations, which are derived from the primary constellation by means of geometric interpolation in the signal space, are employed. It should be noted that for its nature, index modulation in the space domain is not restricted to a MIMO context. Actually, it can be also implemented in a cooperative communication scenario with multiple geometrically distributed relays. For example, each relay may determine its behavior at the next time slot, i.e., whether to transmit or keep idle, by decoding the signal sent by the source node and transmit additional information to the destination node [21].

The above-mentioned schemes perform index modulation at the transmitter side. Actually, the concept of SM is also extendable to the receiver side. In [22], in favor of the zero-forcing (ZF) or minimum mean square error (MMSE) pre-coding, pre-coding SM (PSM) is proposed, which activates a single receive antenna at each time slot and conveys additional information through its index. At the cost of the requirement of CSIT and multiple RF chains, PSM is shown to outperform most transmitter-side SM schemes and the classical MIMO scheme. Attracted by its advantages, plenty of work related to PSM have been also proposed [23–25]. Generalized PSM (GPSM), which allows PSM to activate multiple receive antennas, is designed in [23]. In [24], generalized pre-coding aided quadrature SM (GPQSM) is proposed as another extension of PSM, where the transmitted information is further embedded into the real and imaginary signals. By regarding the geometrically distributed relays as receive antennas, the idea of PSM is applied to a dual-hop relaying, where the source node of multiple antennas employs pre-coding operation to select a single relay, whose index carries information, to forward signals in an SM manner to the destination node [25].

The members of SM family may encounter an identical problem in the implementation, i.e., the number of index combinations is usually not a power of two. To solve this problem, many effective schemes have been designed to, such as fractional bit encoding [26], bit padding [27], and constellation order varying [28]. On the other hand, with CSIT, the capacity or error performance of SM related schemes can be further enhanced by using, e.g., antenna selection [29] and un-equiprobable antenna activation [30]. For other issues regarding space domain index modulation techniques, especially SM and its variants, interested readers may refer to the recent surveys of [31, 32].

1.2.2 Space-Time Domain Index Modulation Overview

Space-time domain index modulation techniques should jointly explore the space and time resources for index modulation. By this definition, the famous space-time shift keying (STSK), Trellis coded spatial modulation (TCSM), space-time block coded spatial modulation (STBC-SM), and differential spatial modulation (DSM) fall within this scope.

STSK first appeared in [33], where one out of Q dispersion space-time matrices is selected for transmission during each communication period, thus conveying additional $\log_2(Q)$ bits through matrix selection to the information bits carried on the modulated symbols. By properly choosing the number of dispersion matrices, the size of the modulation scheme, and the numbers of transmit and receive antennas, STSK achieves a flexible tradeoff between the diversity gain and multiplexing gain. Moreover, unlike the classical SM scheme, which only attains the receive diversity gain, STSK harvests both transmit and receive diversity gains. To further exploit the potential of STSK, the same authors proposed a generalized STSK (GSTSK) scheme, which activates P matrices out of Q dispersion matrices for transmission [34]. The above-mentioned work encounter a common problem, i.e., the optimal ML detection brings about high computational complexity, especially for a high-order modulation or a large number of transmit antennas. To solve this problem, researchers have been focusing on the design of low-complexity detectors for STSK [35–37] and GSTSK [38] systems.

TCSM was first proposed in [39], motivated by the idea of trellis coded modulation (TCM) and SM. In this scheme, the transmit antennas are partitioned into many sub-sets, in which the spacing between antennas is maximized. Later, a new TCSM scheme was proposed in [40] to directly combine TCM and SM, which obtains the time diversity gain in both uncorrelated and correlated channels. Another space-time index modulation scheme, called STBC-SM, first appeared in [41], which takes advantages of both SM and STBC. More extensions of STBC-SM were also studied, such as high-rate STBC-SM [42], STBC-SM with error correcting coding [43], STBC-SM for two-way relaying [44], and multi-strata STBC-SM [45]. Interested readers may refer to [31] for more details.

DSM was introduced first in [46] and later in [47]. In DSM, the information bits are comprised of two parts: The first part selects a space-time matrix indicating the antenna activation order and the second part determines the modulated symbols. Compared with coherently detected SM, DSM bypasses any CSI at the transmitter or receiver, while preserving the single RF chain property at the cost of no more than 3dB performance penalty [48]. The optimal ML detection of DSM, which jointly detects the antenna activation order and modulated symbols, however, leads to prohibitive computational complexity as the number of transmit antennas or modulation order increases. To reduce the detection complexity, many low-complexity detection methods were proposed. By decoupling the detection of antenna activation order and modulated symbols, the authors in [49] proposed a symbol-by-symbol detection method. Inspired by the fact that the antenna activation order has memory, the authors in [50] proposed a Viterbi-like algorithm to achieve a low-complexity search. A sphere decoding idea was also proposed to this end in [51]. To exploit the diversity of DSM, the authors in [52] proposed a cyclic signal constellation to achieve the full diversity for two transmit antennas. The more general design for DSM to achieve the full diversity with an arbitrary number of antennas was proposed in [53]. The performance of DSM depends on the encoding strategy of the antenna activation order. In classical DSM [47], the antenna activation orders are set to follow the lexicographic order, which may lead to huge bit difference between two similar orders. Noticing this problem, the authors in [54] proposed gray coded antenna activation order to further improve the diversity gain and coding gain. More recently, combined with transmitter pre-coding, pre-coding aided DSM (PDSM) was proposed in [55] to map the information bits to the receiver side. DSM can be also applied to different communication scenarios, such as the underwater acoustic (UWA) communication system [56], the full-duplex system [57], and the cooperative communication system [58]. Explicitly, the milestones of DSM are summarized in Table. 1.1.

1.2.3 Frequency Domain Index Modulation Overview

Index modulation can be also performed in the frequency domain. The medium used to this end is usually either the frequency carriers or orthogonal frequency division multiplexing (OFDM) subcarriers. Frequency quadrature amplitude modulation (FQAM) is one of the famous realizations that modulate signals via different frequency carriers [59]. FQAM combines frequency shift keying (FSK) and quadrature amplitude modulation (QAM), providing a desired tradeoff between the transmit power and the spectral efficiency of the system. A great deal of research has been focused on FQAM ever since its emergence and the most recent effort shows that

Table 1.1 History of DSM

Year	Authors	Contributions
2013	Bian et al. [46]	First proposed DSM, which completely dispenses with any CSI at the transmitter or receiver
2014	Wen et al. [48]	Evaluated DSM performance with 2 transmit antennas
	Li et al. [56]	Applied DSM to UWA communications
2015	Bian et al. [47]	Further investigated DSM and showed more results
	Xiao et al. [49]	Proposed a low-complexity detector for DSM, which operates on a symbol-by-symbol basis
	Wen et al. [50]	Proposed a near-ML low-complexity detector for DSM, which resorts to the Viterbi decoding idea
	Li et al. [51]	Proposed a detector for DSM using sphere decoding
	Zhang et al. [52]	Exploited two-transmit-diversity gain for DSM
	Rajashekar et al. [53]	Proposed a method to achieve full diversity for DSM with arbitrary number of transmit antennas
	Zhang et al. [55]	Proposed a pre-coding aided DSM (PDSM) scheme
	Xu et al. [57]	Proposed a self-interference cancellation scheme for full-duplex systems using the idea of DSM
2016	Li et al. [54]	Encoded antenna activation orders with Gray codes
	Zhang et al. [58]	Proposed a dual-hop hybrid DSM (DH-DSM) scheme

by employing FQAM in a downlink OFDM access (OFDMA) system, the inter-cell interference is not Gaussian and the channel capacity can be further increased [60].

As SM becomes more and more popular, researchers begin to extend the principle of SM to OFDM subcarriers. The first attempt was called subcarrier index modulation OFDM (SIM-OFDM) [61]. In SIM-OFDM, the bit stream is split into two sub-streams: The first one is of length N and the second one is of length $N/2\log_2(M)$, where N is the number of total OFDM subcarriers and M denotes the cardinality of the signal constellation. Each bit of the first sub-stream is assigned to a subcarrier and the subcarriers associated with the majority bit value are activated for data transmission. To avoid ambiguity, half subcarriers transmit $N/2$ modulated symbols and the remaining active subcarriers are dedicated to signaling. However, SIM-OFDM assumes perfect detection of the signaling at the receiver and an incorrect detection of a subcarrier state may lead to the incorrect demodulation of all subsequent modulated symbols. To avoid this error propagation problem, an enhanced SIM-OFDM (ESIM-OFDM) scheme was later proposed in [62]. Unlike SIM-OFDM, ESIM-OFDM uses one bit to control two adjacent subcarriers such that only one subcarrier is activated at a time. Therefore, for an OFDM symbol, $N/2$ bits can be conveyed from subcarrier activation. At the receiver, the detection of the state of any subcarrier can be easily performed by comparing the power with that of its adjacent subcarrier. However, the spectral efficiency of ESIM-OFDM tends to be much smaller than that of plain OFDM when a high-order modulation is assumed for the symbols carried on the active subcarriers since the number of active subcarriers is fixed. The authors in [62] raised the idea of relaxing this limitation by allowing activation of more subcarriers. Unfortunately, no further attempt at practical implementation has been found.

The aforementioned incomplete work in [62] was carried out independently in [63, 64], where the new scheme is named OFDM with index modulation (OFDM-IM). It should be noted that the same idea appeared earlier in 1999 [65]. However, the emergence of OFDM-IM brings to the public an entirely new horizon from the concept of index modulation.

In OFDM-IM, the selection of m inactive subcarriers from L subcarriers according to the incoming bits is not a trial problem especially when L is large. So far, there have been two different efficient solutions available in the literature: One is nature binary code [65] and the other is based on combinatorial number theory [64], both exhibiting the same complexity. On the other hand, the optimal ML detection of OFDM-IM signals becomes quite involved since it jointly considers all possible subcarrier activation patterns (SAPs) and signal constellation points. To solve this problem, the authors in [64] proposed a log-likelihood ratio (LLR) detector as an alternative, achieving not only near ML performance but also identical computational complexity to that of the plain OFDM detector. The LLR detector, however, necessitates a prior knowledge of the noise variance. Fortunately, the later proposed near-ML detector with the same computational complexity successfully circumvents this problem [66, 67]. Thanks to the above efforts, the implementation of OFDM-IM becomes practical.

The theoretical analysis of the uncoded bit error rate (BER) reveals that OFDM-IM outperforms plain OFDM [64]. This advantage and the nature of low PAPR have motivated a great deal of follow-up research more recently. A rigorous study from the perspective of coded performance is conducted in [68, 69], where results show that (1) OFDM-IM achieves the maximum rate if and only if the subcarriers within each group experience independent fading, which can be ensured in practice by applying interleaved grouping; (2) The advantage of OFDM-IM over plain OFDM can be maximized by choosing a specific number of inactive subcarriers, i.e., m, typically 1 or 2, for a given signal-to-noise ratio (SNR), and is more noticeable under PSK input. The energy efficiency of OFDM-IM is studied in [70], where superior performance is revealed. A tight upper bound on the BER of OFDM-IM systems is provided in [71]. The BER performance of OFDM-IM systems in the presence of carrier frequency offset is investigated in [72]. The potential of OFDM-IM in ICI mitigation due to the presence of empty subcarriers is explored in [73] and its adaptation to the environment of high mobility, such as vehicular-to-X (V2X) and UWA communications, is verified [68, 74, 75]. Based on the minimum Euclidean distance criterion, a subcarrier allocation scheme assuming the CSIT is tailed to OFDM-IM [76]. To avoid detecting the invalid SAPs, a fractional bit encoding scheme is designed in [77]. In [78], the transmit diversity of OFDM-IM is improved from unity to two by considering the coordinate interleaved orthogonal designs (CIODs) principle. Two variable OFDM-IM systems are proposed in [79], including OFDM with generalized index modulation (OFDM-GIM) that allows a higher number of possible SAPs to be encoded, and OFDM with in-phase/quadrature index modulation (OFDM-I/Q-IM) that expands the index domain to include both the in-phase and quadrature dimensions [67]. In [80], OFDM-IM is further combined with the MIMO transmission technique to form MIMO-OFDM-IM.

1.3 Organization of the Monograph

The organization of this monograph is described as follows. The representative space domain index modulation techniques, namely SM and its variants, are introduced in Chap. 2. The representative space-time domain index modulation techniques, namely DSM and its variants, are presented in Chap. 3. The representative frequency-domain index modulation techniques, namely OFDM-IM and its variants, are studied in Chap. 4. A summary and discussion of future works are presented in Chap. 5.

References

1. V. Tarokh, H. Jafarkhani, A.R. Calderbank, Space-time block codes from orthogonal designs. IEEE Trans. Inform. Theory **45**, 1456–1467 (1999)
2. V. Tarokh, N. Seshadri, A.R. Calderbank, Space-time codes for high data rate wireless communication: performance criterion and code construction. IEEE Trans. Inform. Theory **44**, 744–765 (1998)
3. P. Wolniansky, G. Foschini, G. Golden, R. Valenzuela, V-BLAST: an architecture for realizing very high data rates over the rich-scattering wireless channel, in *Proceedings of Int'l Symp. Signals, Syst., Electron.*, Pisa, Italy, 1998, pp. 295–300
4. J.G. Andrews, S. Buzzi, W. Choi, S.V. Hanly, A. Lozano, A.C.K. Soong, J.C. Zhang, What will 5G be? IEEE J. Sel. Area Commun. **32**(6), 1065–1082 (2014)
5. T.C. Group, Smart 2020: enabling the low carbon economy in the information age (2008)
6. C.-X. Wang, X. Gao, X. You, Y. Yang, D. Yuan, H.M. Aggoune, H. Haas, S. Fletcher, E. Hepsaydir, Cellular architecture and key technologies for 5G wireless communication networks. IEEE Commun. Mag. **52**(2), 122–130 (2014)
7. D.A. Basnayaka, M. Di Renzo, H. Haas, Massive but few active MIMO. IEEE Tran. Veh. Technol. **65**(9), 6861–6877 (2016)
8. Y. Chau, S.CH. Yu, Space modulation on wireless fading channels. Proc. IEEE Veh. Technol. Conf. Fall **3**, 1668–1671 (2001)
9. H. Haas, E. Costa, E. Schultz, Increasing spectral efficiency by data multiplexing using antennas arrays. Proc. IEEE Int. Symp. Personal, Indoor, Mobile Radio Commun. **2**, 610–613 (2002)
10. S. Song, Y. Yang, Q. Xiong, K. Xie, B. Jeong, B. Jiao, A channel hopping technique I: theoretical studies on band efficiency and capacity. Proc. IEEE Int. Conf. Commun., Circuits Syst. **1**, 229–233 (2004)
11. Y. Yang, B. Jiao, Information-guided channel-hopping for high data rate wireless communication. IEEE Commun. Lett. **12**(4), 225–227 (2008)
12. R. Mesleh, H. Haas, S. Sinanovic, C. Ahn, S. Yun, Spatial modulation. IEEE Trans. Veh. Tech. **57**(4), 2228–2242 (2008)
13. J. Jeganathan, A. Ghrayeb, L. Szczecinski, Spatial modulation: optimal detection and performance analysis. IEEE Commun. Lett. **12**(8), 545–547 (2008)
14. J. Jeganathan, A. Ghrayeb, L. Szczecinski, A. Ceron, Space shift keying modulation for MIMO channels. IEEE Trans. Wirel. Commun. **8**(7), 3692–3703 (2009)
15. M. Renzo, H. Haas, P. Grant, Spatial modulation for multiple-antenna wireless systems: a survey. IEEE Commun. Mag. **49**(12), 182–191 (2011)
16. J. Jeganathan, A. Ghrayeb, L. Szczecinski, Generalized space shift keying modulation for MIMO Channels, in *Proceedings of IEEE Symposium on Personal, Indoor and Mobile Radio Communications*, French Riviera, France, Sept. 2008, pp. 1–5
17. A. Younis, N. Serafimovski, R. Mesleh, H. Haas, Generalised spatial modulation, in *Proceedings if 2010 Signals, Systems, and Computers*, pp. 1498–1502

18. J. Wang, S. Jia, J. Song, Generalised spatial modulation system with multiple active transmit antennas and low complexity detection scheme. IEEE Trans. Wirel. Commun. **11**(4), 1605–1615 (2012)
19. R. Mesleh, S. Ikki, H.M. Aggoune, Quadrature spatial modulation. IEEE Trans. Veh. Technol. **64**(6), 2738–2742 (2015)
20. C.-C. Cheng, H. Sari, S. Sezginer, Y. Su, Enhanced spatial modulation with multiple signal constellations. IEEE Trans. Commun. **63**(6), 2237–2248 (2015)
21. Y. Yang, S. Aissa, Information-guided transmission in decode-and-forward relaying systems: spatial exploitation and throughput enhancement. IEEE Trans. Wirel. Commun. **10**(7), 2341–2351 (2011)
22. L. Yang, Transmitter preprocessing aided spatial modulation for multiple-input multiple-output systems, in *Proceedings of 73rd IEEE Vehicular Technology Conference (VTC Spring)*, May 2011, pp. 1–5
23. R. Zhang, L. Yang, L. Hanzo, Generalised pre-coding aided spatial modulation. IEEE Trans. Wirel. Commun. **12**(11), 5434–5443 (2013)
24. J. Li, M. Wen, X. Cheng, Y. Yan, S. Song, M.H. Lee, Generalised pre-coding aided quadrature spatial modulation. IEEE Trans. Veh. Technol. (2016, in press)
25. A. Stavridis, D. Basnayaka, S. Sinanovic, M. Di Renzo, H. Haas, A virtual MIMO dual-hop architecture based on hybrid spatial modulation. IEEE Trans. Commun. **62**(9), 3161–3179 (2014)
26. N. Serafimovski, M. Renzo, S. Sinanovic, R. Mesleh, H. Haas, Fractional bit encoded spatial modulation (FBE-SM). IEEE Commun. Lett. **14**(5), 429–431 (2010)
27. Y. Yang, S. Aissa, Bit-padding information guided channel hopping. IEEE Commun. Lett. **15**(2), 163–165 (2011)
28. Y. Yang, S. Aissa, Information guided channel hopping with an arbitrary number of transmit antennas. IEEE Commun. Lett. **16**(10), 1552–1555 (2012)
29. R. Rajashekar, K.V.S. Hari, L. Hanzo, Antenna selection in spatial modulation systems. IEEE Commun. Lett. **17**(3), 521–524 (2013)
30. C. Liu, M. Ma, Y. Yang, B. Jiao, Optimal spatial-domain design for spatial modulation capacity maximization. IEEE Commun. Lett. **20**(6), 1092–1095 (2013)
31. M. Renzo, H. Haas, A. Ghrayeb, S. Sugiura, L. Hanzo, Spatial modulation for generalized MIMO: challenges, opportunities and implementation. Proc. IEEE **102**(1), 56–103 (2014)
32. P. Yang, M. Di Renzo, Y. Xiao, S. Li, L. Hanzo, Design guidelines for spatial modulation. IEEE Commun. Surv. Tuts. **17**(1), 6–26 (2015)
33. S. Sugiura, S. Chen, L. Hanzo, Coherent and differential space-time shift keying: a dispersion matrix approach. IEEE Trans. Commun. **58**(11), 3219–3230 (2010)
34. S. Sugiura, S. Chen, L. Hanzo, Generalized space-time shift keying designed for flexible diversity-, multiplexing- and complexity-tradeoffs. IEEE Trans. Wirel. Commun. **10**(4), 1144–1153 (2011)
35. C. Xu, S. Sugiura, S.X. Ng, Lajos Hanzo, Reduced-complexity soft-decision aided space-time shift keying. IEEE Signal Process. Lett. **18**(10), 547–550 (2011)
36. S. Sugiura, C. Xu, S.X. Ng, L. Hanzo, Reduced-complexity coherent versus non-coherent QAM-aided space-time shift keying. IEEE Trans. Commun. **59**(11), 3090–3101 (2011)
37. P. Yang, Y. Xiao, L. Li, Q. Tang, S. Li, An improved matched-filter based detection algorithm for space-time shift keying systems. IEEE Signal Process. Lett. **19**(5), 271–274 (2012)
38. S. Sugiura, C. Xu, S.X. Ng, L. Hanzo, Reduced-complexity iterative-detection-aided generalized space-time shift keying. IEEE Trans. Veh. Technol. **61**(8), 3656–3664 (2012)
39. R. Mesleh, M.D. Renzo, H. Haas, P.M. Grant, Trellis coded spatial modulation. IEEE Trans. Wirel. Commun. **9**(7), 2349–2361 (2010)
40. E. Basar, U. Aygolu, E. Panayirci, H.V. Poor, New trellis code design for spatial modulation. IEEE Trans. Wirel. Commun. **10**(8), 2670–2680 (2011)
41. E. Basar, U. Aygolu, E. Panayirci, H.V. Poor, Space-time block coded spatial modulation. IEEE Trans. Commun. **59**(3), 823–832 (2011)

42. X. Li, L. Wang, High rate space-time block coded spatial modulation with cyclic structure. IEEE Commun. Lett. **18**(4), 532–535 (2014)
43. L. Wang, Z. Chen, X. Wang, A space-time block coded spatial modulation from (n, k) error correcting code. IEEE Wirel. Commun. Lett. **3**(1), 54–57 (2014)
44. K.G. Unnikrishnan, B.S. Rajan, Space-time coded spatial modulated physical layer network coding for two-way relaying. IEEE Trans. Commun. **14**(1), 331–342 (2015)
45. C. Jeon, J. Lee, Multi-strata space-time coded spatial modulation. IEEE Commun. Lett. **19**(11), 1945–1948 (2015)
46. Y. Bian, M. Wen, X. Cheng, H.V. Poor, B. Jiao, A differential scheme for spatial modulation, in *Proceedings of IEEE Global Communications Conference* Atlanta, USA, Dec 2013
47. Y. Bian, X. Cheng, M. Wen, L. Yang, H.V. Poor, B. Jiao, Differential spatial modulation. IEEE Trans. Veh. Technol. **64**(7), 3262–3268 (2015)
48. M. Wen, Z. Ding, X. Cheng, Y. Bian, Performance analysis of Differential spatial modulation with two transmit antennas. IEEE Commun. Lett. **18**(3), 475–478 (2014)
49. L. Xiao, P. Yang, X. Lei, Y. Xiao, S. Fan, S. Li, W. Xiang, A low-complexity detection scheme for differential spatial modulation. IEEE Commun. Lett. **19**(9), 1516–1519 (2015)
50. M. Wen, X. Cheng, Y. Bian, H.V. Poor, A low-complexity near-ML differential spatial modulation detector. IEEE Signal Process. Lett. **22**(11), 1834–1838 (2015)
51. Z. Li, X. Cheng, S. Han, M. Wen, L. Yang, B. Jiao, A low-complexity optimal sphere decoder for differential spatial modulation, in *Proceedings of IEEE Global Communications Conference* San Diego, USA, Dec 2015
52. W. Zhang, Q. Yin, H. Deng, Differential full diversity spatial modulation and its performance analysis with two transmit antennas. IEEE Commun. Lett. **19**(4), 677–680 (2015)
53. R. Rajashekar, N. Ishikawa, S. Sugiura, K.V.S. Hari, L. Hanzo, Full-diversity dispersion matrices from algebraic field extensions for differential spatial modulation. IEEE Trans. Veh. Technol. (2016, in press)
54. J. Li, M. Wen, X. Cheng, Y. Yan, S. Song, M.H. Lee, Differential spatial modulation with Gray coded antenna activation order. IEEE Commun. Lett. **20**(6), 1100–1103 (2016)
55. M. Zhang, M. Wen, X. Cheng, L. Yang, Pre-coding aided differential spatial modulation, in *Proceedings of IEEE Global Communications Conference*, Dec. 2015, pp. 1–6
56. Z. Li, X. Cheng, M. Wen, L. Yang, B. Jiao, Differential spatial modulated OFDM in underwater acoustic communications, in *Proceedings of IEEE International Conference on Signal Processing, Communications and Computing (ICSPCC)*, Guilin, China, Aug. 2014, pp. 913–918
57. M. Xu, M. Wen, Y. Feng, F. Ji, W. Pan, A novel self-interference cancellation scheme for full duplex with differential spatial modulation, in *Proceedings of IEEE 26th Annual International Symposium on Personal, Indoor, and Mobile Radio Communications (PIMRC 2015)*, Hong Kong. China, Sept. 2015, pp. 482–486
58. M. Zhang, M. Wen, X. Cheng, L. Yang, A dual-hop virtual MIMO architecture based on hybrid differential spatial modulation. IEEE Trans. Wirel. Commun. **15**(9), 6356–6370 (2016)
59. R. Padovani, J. Wolf, Coded phase/frequency modulation. IEEE Trans. Commun. **COM-34**(5) (1986)
60. S. Hong, M. Sagong, C. Lim, S. Cho, K. Cheun, K. Yang, Frequency and quadrature-amplitude modulation for downlink cellular OFDMA networks. IEEE J. Sel. Area Commun. **32**(6), 1256–1267 (2014)
61. R.A. Alhiga, H. Haas, Subcarrier-index modulation OFDM, in *Proceedings of IEEE 20th International Symposium on Personal, Indoor and Mobile Radio Communications (PIMRC)*, Tokyo, Japan, Sept. 2009, pp. 177–181
62. D. Tsonev, S. Sinanovic, H. Haas, Enhanced subcarrier index modulation (SIM) OFDM, in *Proceedings of IEEE Global Communications Conference (GLOBECOM) Workshops*, Houston, TX, USA, Dec. 2011, pp. 728–732
63. E. Basar, U. Aygolu, E. Panayirci, H.V. Poor, Orthogonal frequency division multiplexing with index modulation, in *Proceedings of IEEE Global Communications Conference (GLOBECOM)*, Anahem, CA, USA, Dec. 2012, pp. 4741–4746

64. E. Basar, U. Aygolu, E. Panayirci, H.V. Poor, Orthogonal frequency division multiplexing with index modulation. IEEE Trans. Signal Process. **61**(22), 5536–5549 (2013)
65. P.K. Frenger, N.A.B. Svensson, Parallel combinatory OFDM signaling. IEEE Trans. Commun. **47**(4), 558–567 (1999)
66. M. Wen, Y. Li, X. Cheng, L. Yang, Index modulated OFDM with ICI self-cancellation in underwater acoustic communications, in *Proceedings of IEEE Asilomar Conference on Signals, Systems, and Computers*, Pacific Grove, USA, Nov. 2014, pp. 338–342
67. B. Zheng, F. Chen, M. Wen, F. Ji, H. Yu, Y. Liu, Low-complexity ML detector and performance analysis for OFDM with in-phase/quadrature index modulation. IEEE Commun. Lett. **19**(11), 1893–1896 (2015)
68. X. Cheng, M. Wen, L. Yang, Y. Li, Index modulated OFDM with interleaved grouping for V2X communications, in *Proceedings of IEEE ITSC*, Qingdao, China, Oct. 2014, pp. 1097–1104
69. M. Wen, X. Cheng, M. Ma, B. Jiao, H.V. Poor, On the achievable rate of OFDM with index modulation. IEEE Trans. Signal Process. **64**(8), 1919–1932 (2016)
70. M. Wen, X. Cheng, L. Yang, Optimizing the energy efficiency of OFDM with index modulation, in *Proceedings of IEEE ICCS*, Macau, China, Nov. 2014, pp. 31–35
71. Y. Ko, A tight upper bound on bit error rate of joint OFDM and multi-carrier index keying. IEEE Commun. Lett. **18**(10), 1763–1766 (2014)
72. Q. Ma, P. Yang, Y. Xiao, H. Bai, S. Li, Error probability analysis of OFDM-IM with carrier frequency offset. IEEE Commun. Lett. **20**(12), 2434–2437 (2016)
73. M. Wen, X. Cheng, L. Yang, Y. Li, X. Cheng, F. Ji, Index modulated OFDM for underwater acoustic communications. IEEE Commun. Mag. **54**(5), 132–137 (2016)
74. Y. Li, M. Zhang, X. Cheng, M. Wen, L. Yang, Index modulated OFDM with intercarrier interference cancellation, in *Proceedings of IEEE International Conference on Communications (ICC)*, Kuala Lumpur, Malaysia, May 2016
75. Y. Li, M. Wen, L. Yang, Index modulated OFDM with ICI self-cancellation for V2X communications, in *Proceedings of IEEE International Conference on Computing, Networking and Communications (ICNC 2016)*, Hawaii, HI, USA, Feb. 2016, pp. 1–5
76. Q. Ma, Y. Xiao, L. Dan, P. Yang, L. Peng, S. Li, Subcarrier allocation for OFDM with index modulation. IEEE Commun. Lett. **20**(7), 1469–1472 (2016)
77. A.I. Siddiq, Low complexity OFDM-IM detector by encoding all possible subcarrier activation patterns. IEEE Commun. Lett. **20**(3), 446–449 (2016)
78. E. Basar, OFDM with index modulation using coordinate interleaving. IEEE Wirel. Commun. Lett. **4**(4), 381–384 (2015)
79. R. Fan, Y.J. Yu, Y.L. Guan, Generalization of orthogonal frequency division multiplexing with index modulation. IEEE Trans. Wirel. Commun. **14**(10), 5350–5359 (2015)
80. E. Basar, On multiple-input multiple-output OFDM with index modulation for next generation wireless networks. IEEE Trans. Signal Process. **64**(15), 3868–3878 (2016)

Chapter 2
Space Domain Index Modulation

In this chapter, we discuss the realization of index modulation in the space domain. First, we take a brief review of the pioneering space domain index modulation technique, namely spatial modulation (SM). Then, we propose the generalized pre-coding aided quadrature spatial modulation (GPQSM) scheme, which extends SM to the receiver side, and the virtual spatial modulation (VSM) scheme, which can be used as either a transmitter-side SM or a receiver-side SM scheme depending on the configuration of transceiver antennas. Finally, we discuss the application of space domain index modulation techniques by taking SM as a representative to cooperative communications.

2.1 Spatial Modulation

In this section, we first present the principle of SM and then review some popular low-complexity detection methods. Finally, the performance of SM is evaluated via computer simulations.

2.1.1 System Model

In SM, the information bits are conveyed by not only the modulated symbol but also the index of the active antenna. Let us consider an $N_t \times N_r$ multiple-input multiple-output (MIMO) system with an M-ary constellation $\mathcal{X}$, where N_t, N_r, and M denote the number of transmit antennas, the number of receive antennas, and the cardinality of the constellation, respectively. The transceiver structure of SM is depicted in Fig. 2.1. At each time slot, the incoming $\log_2(N_t M)$ information bits are divided into two parts. The first part of $\log_2(M)$ bits draws a modulated symbol from $\mathcal{X}$. The second part of $\log_2(N_t)$ bits selects an antenna to transmit this symbol. Here, we simply assume that N_t equals a power of two. If this condition cannot be satisfied in practice, we may select partial antennas whose number equals a power of two to

M. Wen et al., *Index Modulation for 5G Wireless Communications*,
Wireless Networks, DOI 10.1007/978-3-319-51355-3_2

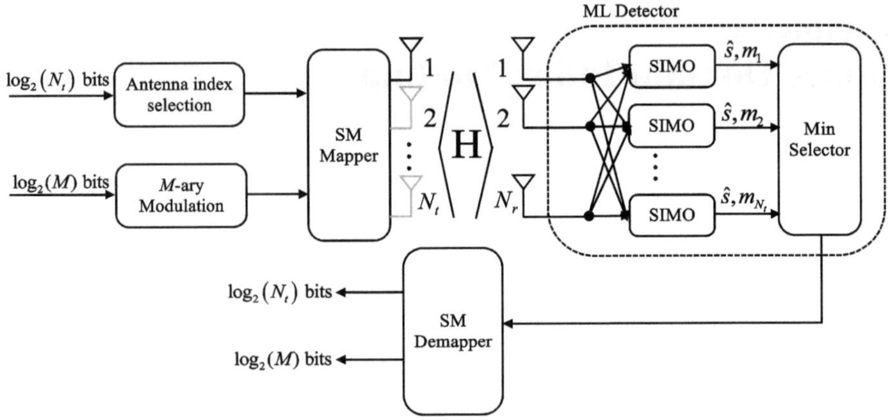

Fig. 2.1 Transceiver structure of SM

perform SM or employ the fractional bit encoding (FBE) to involve all antennas for index modulation [1]. Note that during each transmission only the chosen antenna is active while the other antennas remain idle.

Assume that at a given time slot, $s \in \mathcal{X}$ is selected and transmitted via the k-th antenna. Let $\mathbf{H} \in \mathbb{C}^{N_r \times N_t}$ denote the MIMO channel matrix, whose (i, j)-th element $h_{i,j}$ represents the channel from the j-th transmit antenna to the i-th receive antenna. Then, the received signal vector can be expressed as

$$\mathbf{y} = \mathbf{h}_k s + \mathbf{w}, \tag{2.1}$$

where $\mathbf{h}_k$ denotes the k-th column of $\mathbf{H}$ and $\mathbf{w} \in \mathbb{C}^{N_r \times 1}$ is the additive white Gaussian noise (AWGN) vector.

From (2.1), the maximum-likelihood (ML) detection can be formulated as [2]

$$[\hat{k}, \hat{s}] = \underset{k \in \{1,\ldots,N_t\}, s \in \mathcal{X}}{\arg\min} \|\mathbf{y} - \mathbf{h}_k s\|^2. \tag{2.2}$$

The ML detection in (2.2) is optimal at the expense of high search complexity, which is of order $\mathcal{O}(N_t M)$, since it invokes joint detection of the antenna index and modulated symbol.

2.1.2 *Low-Complexity Detection*

Attentive to the high computational complexity of the ML detection, many researchers have been focusing on the design of low-complexity detection methods for SM. In the following, we will introduce some popular ones.

The pioneering work is presented in [3], where the matched filtering (MF) detection is proposed. In this method, the detections for the antenna index and modulated symbol are decoupled. Specifically, first, the active antenna is estimated, which is realized by

$$\hat{k} = \underset{k\in\{1,\ldots,N_t\}}{\arg\max}\ \hat{y}_k, \tag{2.3}$$

where $\hat{y}_k = |\mathbf{h}_k^H\mathbf{y}|/\|\mathbf{h}_k\|^2$. Then, the modulated symbol is estimated via

$$\hat{s} = \underset{s\in\mathcal{X}}{\arg\min}\ ||\mathbf{y} - \mathbf{h}_{\hat{k}}s||^2. \tag{2.4}$$

The MF detection, however, comes at a cost of system performance, since it is required that $\|\mathbf{h}_j\|^2 < \|\mathbf{h}_k\|^2$ for all $j \neq k$ [2].

Another alternative is the signal vector based detection [4]. The main difference between this method and the MF detection lies in the estimation of the antenna index. In signal vector based detection, the active antenna is determined when the angle between its corresponding channel vector and the received signal vector is the minimum one, i.e.,

$$\begin{aligned}\theta(k) &= \arccos\frac{|\langle\mathbf{h}_k, \mathbf{y}\rangle|}{\|\mathbf{h}_k\|\cdot\|\mathbf{y}\|},\\ \hat{k} &= \underset{k\in\{1,\ldots,N_t\}}{\arg\min}\ \theta(k).\end{aligned} \tag{2.5}$$

However, the inaccuracy of the angle measurement due to the presence of the modulated symbol severely degrades the system performance.

To avoid angle ambiguity caused by the signal vector based detection, the signal vector based list (SVBL) detection is proposed [5]. In SVBL detection, the number of active antenna candidates is relaxed from 1 to L, whose indices, denoted by $\mathcal{L} = \{k_1, \ldots, k_L\}$, are associated with the smallest L angles among $\{\theta(1), \ldots, \theta(N_t)\}$. Then, the antenna index and the modulated symbol are jointly estimated as

$$[\hat{k}, \hat{s}] = \underset{k\in\mathcal{L}, s\in\mathcal{X}}{\arg\min}\ \|\mathbf{y} - \mathbf{h}_k s\|^2. \tag{2.6}$$

Explicitly, the improved performance is achieved at a price of increased complexity, whose amount depends on the value of L.

A similar idea is realized by the distance-based ordered detection [6]. In this method, first, by assuming the activation of the k-th transmit antenna, the modulated symbol carried on is estimated as

$$\hat{r}_k = \underset{s\in\mathcal{X}}{\arg\min}\ ||\mathbf{y} - \mathbf{h}_k s||^2, \tag{2.7}$$

where $k \in \{1, \ldots, N_t\}$. Note that $\{\hat{r}_1, \ldots, \hat{r}_{N_t}\}$ are N_t symbol estimates, but only one of them is necessary. Let d_n denote the distance between r_k and $\hat{r}_k$, given by

$$d_k = ||\mathbf{h}_k|| \, |\hat{r}_k - r_k|. \tag{2.8}$$

Then, by sorting $\{d_1, \ldots, d_{N_t}\}$ in ascending order and obtaining the indices $\mathcal{U} = \{u_1, \ldots, u_p\}$ associated with the first p values, the antenna index and the modulated symbol are jointly estimated as

$$\hat{k} = \arg\min_{k \in \mathcal{U}} \|\mathbf{y} - \mathbf{h}_k \hat{r}_k\|^2, \tag{2.9}$$

$$\hat{s} = \hat{r}_{\hat{k}}, \tag{2.10}$$

where similar to the parameter L in the SVBL detection, the parameter p is used to reach a compromise between the system performance and the computational complexity. Explicitly, as p increases, the distance-based ordered detection approaches the optimal ML detection.

The above-mentioned four detection methods achieve sub-optimal performance. In the following, we introduce two low-complexity optimal detection methods, which reduce the search complexity to an order of $\mathcal{O}(N_t)$. The first one is the hard-limiter based ML detection [7], which applies to a square or rectangular QAM. The idea is motivated by the fact that the M-QAM constellation can be viewed as a product of M_1-PAM and M_2-PAM constellations, where $M = M_1 M_2$. Consequently, assuming the activation of the k-th transmit antenna, the real and the imaginary parts of the modulated symbol carried on can be directly estimated as

$$\Re\{\hat{r}_k\} = \min\left\{\max\left\{2\lfloor\frac{\Re(\hat{y}_k)+1}{2}\rceil, M_1+1\right\}, M_1-1\right\}, \tag{2.11}$$

and

$$\Im\{\hat{r}_k\} = \min\left\{\max\left\{2\lfloor\frac{\Im(\hat{y}_k)+1}{2}\rceil, M_2+1\right\}, M_2-1\right\}, \tag{2.12}$$

respectively, where $\lfloor\cdot\rceil$ denotes the operation of rounding a real number to the nearest integer, $\hat{y}_k$ is given by (2.3), and $k \in \{1, \ldots, N_t\}$. However, only one among $\{\hat{r}_1, \ldots, \hat{r}_{N_t}\}$ is necessary. To determine the necessary one, one can resort to the ML criterion:

$$\hat{k} = \arg\min_{k \in \{1,\ldots,N_t\}} \|\mathbf{y} - \mathbf{h}_k \hat{r}_k\|^2,$$
$$\hat{s} = \hat{r}_{\hat{k}}. \tag{2.13}$$

The second one is the decision-zone based detection [8], which applies to an MPSK input. Assume that $\hat{y}_k = |\hat{y}_k| e^{j\vartheta_k}$ and $\hat{r}_k = e^{j\varphi_k}$, where ϑ_k and φ_k are the phases of

$\hat{y}_k$ and $\hat{r}_k$, respectively. According to the decision zones for the MPSK constellation, the phase of the modulated symbol can be directly estimated as

$$\hat{\varphi}_k = \frac{2\pi}{M} < \lfloor \frac{\vartheta_k M}{2\pi} \rceil >_M, \tag{2.14}$$

where $< \cdot >_M$ denotes the module M operation. Then, having the phase, the symbol candidate can be directly obtained as $\hat{r}_k = e^{j\hat{\varphi}_k}$. Finally, the antenna index and modulated symbol can be estimated via (2.13).

It should be noted that besides all aforementioned methods, we may also apply the ideas of sphere decoding and compressed sensing for detection, which achieve different tradeoffs between the system performance and the computational complexity.

2.1.3 *Performance Evaluation*

The bit error rate (BER) performance of SM is shown in Fig. 2.2, where comparison is made with V-BLAST and their spectral efficiencies are equated to 8 bps/Hz. As can be seen, both V-BLAST and SM achieve the same diversity order of N_r, and SM can outperform V-BLAST by leveraging the numbers of space bits and modulation bits.

Taking the SVBL detector as a representative for all low-complexity SM detectors, we compare its performance with that of the optimal ML detector in Fig. 2.3. Two system configurations are considered, which are $\{N_t = 4, N_r = 2\}$ and $\{N_t = 8, N_r = 4\}$. Recall that for $L = 1$, the SVBL detector degenerates into the signal vector based detector and for $L = N_t$, the SVBL detector becomes the optimal ML detector. From the figure, we observe that the SVBL detector with $L = 1$ exhibits unsatisfactory performance, but the performance is improved significantly as L increases. Interestingly, for a value of L much smaller than N_t, near-optimal performance is already achieved by the SVBL detector.

2.2 Pre-coding Aided Spatial Modulation

In this section, we propose a novel scheme, called GPQSM, which extends SM to the receiver side. In GPQSM, index modulation works on both the in-phase and quadrature parts of the received signals. GPQSM is general and can degenerate into the conventional MIMO scheme. A closed-form upper bound on the average bit error probability (ABEP) of GPQSM is also derived.

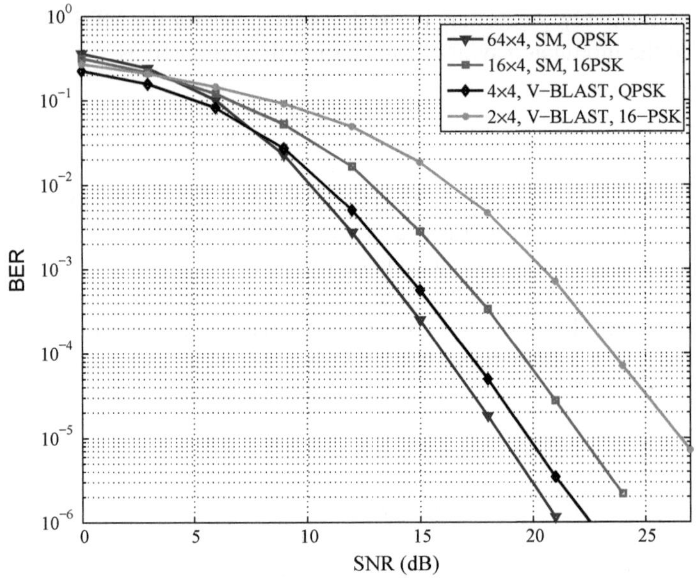

Fig. 2.2 Performance comparison between SM and V-BLAST in terms of BER under the spectral efficiency of 8 bps/Hz

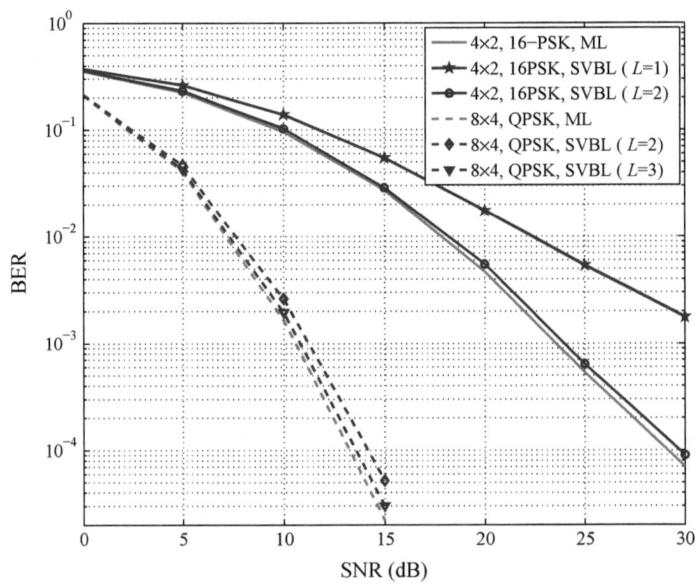

Fig. 2.3 Performance of the SVBL detector

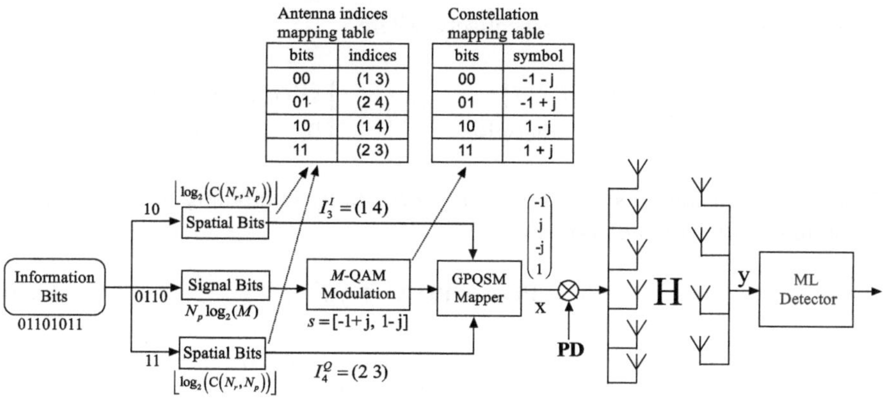

Fig. 2.4 Transceiver structure of GPQSM with $N_t = 6$, $N_r = 4$, and 4-QAM

2.2.1 System Model

The transceiver structure of GPQSM with $N_t = 6$, $N_r = 4$, and 4-QAM is depicted in Fig. 2.4. We always assume $N_t \geq N_r$ and perfect channel state information at the transmitter (CSIT) for ensuring the pre-coding design. In GPQSM, N_p $(1 \leq N_p \leq N_r)$ receive antennas are activated to transmit the real (imaginary) parts of transmitted symbols. There are in total $\mathrm{C}(N_r, N_p)$ possible antenna combinations. However, in order to modulate the space bits, only $N = 2^{\lfloor \log_2(\mathrm{C}(N_r, N_p)) \rfloor}$ combinations will be used. Note that the conventional MIMO scheme is a special case of GPQSM with $N_p = N_r$.

At each time slot, $n = N_p \log_2(M) + 2\lfloor \log_2(\mathrm{C}(N_r, N_p)) \rfloor$ information bits are to be transmitted. First, the $N_p \log_2(M)$ bits are mapped into N_p modulated symbols $\mathbf{s} = \{s(1), \ldots, s(N_p)\}$, of which each can be expressed by $s(\eta) = s^I(\eta) + js^Q(\eta)$, $s(\eta) \in \mathcal{X}$, where $s^I(\eta)$ and $s^Q(\eta)$ represent the in-phase and quadrature parts of $s(\eta)$, respectively. Note that $s^I(\eta) \in \mathcal{X}^I$ and $s^Q(\eta) \in \mathcal{X}^Q$, where $\mathcal{X}^I$ and $\mathcal{X}^Q$ denote the in-phase and quadrature sets of $\mathcal{X}$, respectively. Then, $\lfloor \log_2(\mathrm{C}(N_r, N_p)) \rfloor$ bits are used to select an antenna combination for transmitting $\{s^I(1), \ldots, s^I(N_p)\}$ and the other $\lfloor \log_2(\mathrm{C}(N_r, N_p)) \rfloor$ bits are used to select another antenna combination for transmitting $\{s^Q(1), \ldots, s^Q(N_p)\}$. Denote the ν-th legitimate active antenna combination for in-phase/quadrature parts as

$$
\begin{aligned}
I_\nu^I &= \{i_\nu^I(1), \ldots, i_\nu^I(N_p)\}, \\
I_\nu^Q &= \{i_\nu^Q(1), \ldots, i_\nu^Q(N_p)\},
\end{aligned}
\tag{2.15}
$$

where $i_\nu^I(a), i_\nu^Q(a) \in \{1, \ldots, N_r\}$ with $\nu = 1, \ldots, N$ and $a = 1, \ldots, N_p$. If the u-th and o-th active antenna combinations are chosen by the incoming bits, respectively, then the signal vectors before pre-coding corresponding to the in-phase and quadrature parts can be written as $\mathbf{x}^I = [x^I(1), \ldots, x^I(N_r)]^T$ and

$\mathbf{x}^Q = [x^Q(1), \ldots, x^Q(N_r)]^T$, respectively, where

$$x^I(\eta) = \begin{cases} s^I(\eta), & \eta \in I_u^I \\ 0, & \text{otherwise} \end{cases}, \quad x^Q(\eta) = \begin{cases} s^Q(\eta), & \eta \in I_o^Q \\ 0, & \text{otherwise} \end{cases}.$$

Finally, the transmitted signal vector is generated by combining the in-phase and quadrature parts with pre-coding:

$$\mathbf{X} = \mathbf{PDx} = \mathbf{PD}(\mathbf{x}^I + j\mathbf{x}^Q), \tag{2.16}$$

where $\mathbf{P} \in \mathbb{C}^{N_t \times N_r}$ is the pre-coding matrix and $\mathbf{D}$ is an $N_r \times N_r$ diagonal matrix whose η-th diagonal element is $d(\eta) = \sqrt{1/||\mathbf{p}_\eta||^2}$ with $\mathbf{p}_\eta$ denoting the η-th column of $\mathbf{P}$.

Let $\mathbf{H}$ denote the MIMO channel, whose entries follow the Nakagami-m distribution with parameters $(1, m)$. The received signal vector $\mathbf{y} \in \mathbb{C}^{N_r \times 1}$ can be expressed by

$$\mathbf{y} = \mathbf{HPDx} + \mathbf{w}, \tag{2.17}$$

where $\mathbf{w} \in \mathbb{C}^{N_r \times 1}$ is an AWGN vector with zero mean and covariance matrix $N_P\sigma^2/N_r\mathbf{I}_{N_r}$. It should be noted that the imperfect channel estimation and pre-coding will break the orthogonality between the in-phase and quadrature components, which seriously deteriorates the performance of GPQSM. To ease the analysis, we simply assume perfect CSIT. The linear pre-coding schemes, such as zero-forcing (ZF) and minimum mean squared error (MMSE) pre-coding, can be applied to this system. Here, we only consider ZF pre-coding, which implies that the adopted pre-coder is $\mathbf{P} = \mathbf{H}^H(\mathbf{HH}^H)^{-1}$. In this case, $d(\eta)$ can be expressed by

$$d(\eta) = \frac{1}{\sqrt{[(\mathbf{HH}^H)^{-1}]_{\eta,\eta}}}. \tag{2.18}$$

Therefore, the received signal vector can be rewritten from (2.17) as

$$\mathbf{y} = \mathbf{Dx} + \mathbf{w} = \mathbf{Dx}^I + j\mathbf{Dx}^Q + \mathbf{w}. \tag{2.19}$$

The optimal ML detector is given by

$$\begin{aligned} \hat{\mathbf{x}} &= \arg\min_{\mathbf{x}} \|\mathbf{y} - \mathbf{Dx}\|^2 \\ &= \arg\min_{\mathbf{x}^I, \mathbf{x}^Q} \|\mathbf{y} - (\mathbf{Dx}^I + j\mathbf{Dx}^Q)\|^2. \end{aligned} \tag{2.20}$$

From (2.20), it can be seen that the optimal ML detector requires an exhausting search over all possible transmitted signal vectors. To reduce the search complexity, we propose a low-complexity detection method as follows. Due to the independence between the in-phase and quadrature parts, (2.20) is equivalent to

$$\hat{\mathbf{x}} = \arg\min_{\mathbf{x}^I} \|\mathbf{y}^I - \mathbf{D}\mathbf{x}^I\|^2 + j \arg\min_{\mathbf{x}^Q} \|\mathbf{y}^Q - \mathbf{D}\mathbf{x}^Q\|^2, \tag{2.21}$$

where $\mathbf{y}^I$ and $\mathbf{y}^Q$ represent the in-phase and quadrature parts of $\mathbf{y}$, respectively. Since the detection procedures for the in-phase and quadrature parts are the same, we now only consider the detection for the in-phase part, which can be expressed by

$$\begin{aligned}\hat{\mathbf{x}}^I &= \arg\min_{\mathbf{x}^I} \|\mathbf{y}^I - \mathbf{D}\mathbf{x}^I\|^2 \\ &= \arg\min_{\{x^I(\eta)\}_{\eta=1}^{N_r}} \sum_{\eta=1}^{N_r} |y^I(\eta) - d(\eta)x^I(\eta)|^2. \end{aligned}\tag{2.22}$$

From (2.22), it reveals that the search complexity is of order $(M^I)^{N_p}N$, where M^I denotes the cardinality of $\mathcal{X}^I$. A further reduction of search complexity is possible if we expand (2.22) as

$$\begin{aligned}
&\arg\min_{\{x^I(\eta)\}_{\eta=1}^{N_r}} \sum_{\eta=1}^{N_r} |y^I(\eta) - d(\eta)x^I(\eta)|^2 \\
&= \arg\min_{\{x^I(\eta)\}_{\eta=1}^{N_r}} \sum_{\eta=1}^{N_r} \left[\left(y^I(\eta)\right)^2 - 2d(\eta)x^I(\eta)y^I(\eta) + \left(d(\eta)x^I(\eta)\right)^2\right] \\
&= \arg\min_{I_\nu^I, \{s^I(\eta)\}_{\eta\in I_\nu^I}} \left[\sum_{\eta\in I_\nu^I} \left[\left(d(\eta)s^I(\eta)\right)^2 - 2d(\eta)s^I(\eta)y^I(\eta)\right] + \sum_{\eta=1}^{N_r} \left(y^I(\eta)\right)^2\right] \\
&= \arg\min_{I_\nu^I} \left[\sum_{\eta\in I_\nu^I} \min_{s^I(\eta)} \left[\left(d(\eta)s^I(\eta)\right)^2 - 2d(\eta)s^I(\eta)y^I(\eta)\right]\right].
\end{aligned}\tag{2.23}$$

Define $D(\eta) = \left(d(\eta)\hat{s}^I(\eta)\right)^2 - 2d(\eta)\hat{s}^I(\eta)y^I(\eta)$, where

$$\hat{s}^I(\eta) = \arg\min_{s\in\mathcal{X}^I} |y^I(\eta) - d(\eta)s|^2. \tag{2.24}$$

Therefore, the detection for the active antenna combination in (2.23) is equivalent to

$$\hat{I}_\nu^I = \arg\min_{I_\nu^I} \sum_{\eta\in I_\nu^I} D(\eta). \tag{2.25}$$

It is clear that the order of the search complexity in (2.25) is reduced to $N_r M^I + N$. However, the search complexity is still high especially when N_r or N_p is large. For practical use, it is advisable to first obtain $[z_1, \ldots, z_{N_r}] = \mathrm{sort}(\{D(\eta)\}_{\eta=1}^{N_r})$, where $\mathrm{sort}(\cdot)$ defines an ordering function that reorders the elements of the input vector in ascending order, and z_1, z_{N_r} are the indices of the maximum and minimum values in $\{D(\eta)\}_{\eta=1}^{N_r}$, respectively. Then, we simply obtain the active antenna indices by $\hat{I}_\nu^I = \{z_1, \ldots, z_{N_p}\}$. Based on this manner, we may obtain an illegitimate antenna combination, which does not exist in the mapping table. To relieve this adverse effect, when this event occurs, we select $\hat{I}_\nu^I = \{z_1, \ldots, z_{N_{p-1}}, z_{N_{p+1}}\}$ instead. After getting $\hat{I}_\nu^I$, the estimation of $\mathbf{x}^I$ is directly obtained by

$$\hat{x}^I(\eta) = \begin{cases} \hat{s}^I(\eta), & \eta \in \hat{I}_\nu^I \\ 0, & \text{otherwise} \end{cases}. \tag{2.26}$$

From the proposed detector, it follows that the equivalent detection only leads to a search complexity of order $N_r M^I$. Compared with the ML detector in (2.22), the search complexity is largely reduced.

2.2.2 *Average Bit Error Probability Analysis*

According to (2.16), it indicates that the n information bits can be treated as two independent bits $n_1 = \log_2(M) + \lfloor \log_2(\mathrm{C}(N_r, N_p)) \rfloor$ and $n_2 = \log_2(M) + \lfloor \log_2(\mathrm{C}(N_r, N_p)) \rfloor$ corresponding to in-phase and quadrature parts, respectively. Without loss of generality, we only analyze the average bit error probability (ABEP) for the n_1 bits.

An upper bound on ABEP for the n_1 bits can be derived according to the union bounding technique as [9]

$$P_E^I \leq \frac{1}{n_1 2^{n_1}} \sum_{\mathbf{x}^I} \sum_{\hat{\mathbf{x}}^I \neq \mathbf{x}^I} N(\mathbf{x}^I, \hat{\mathbf{x}}^I) \Pr(\mathbf{x}^I \to \hat{\mathbf{x}}^I), \tag{2.27}$$

where $\Pr(\mathbf{x}^I \to \hat{\mathbf{x}}^I)$ denotes the pairwise error probability (PEP) of detecting $\hat{\mathbf{x}}^I$ when $\mathbf{x}^I$ is transmitted and $N(\mathbf{x}^I, \hat{\mathbf{x}}^I)$ measures the number of bits in difference between $\mathbf{x}^I$ and $\hat{\mathbf{x}}^I$.

The PEP conditioned on $\mathbf{D}$ is given by

$$\Pr(\mathbf{x}^I \to \hat{\mathbf{x}}^I | \mathbf{D}) = Q\left(\sqrt{\frac{\lambda_I}{2}\gamma}\right), \tag{2.28}$$

where $\lambda_I = \mathbf{c}^H\mathbf{D}^2\mathbf{c}$ with $\mathbf{c} = \mathbf{x}^I - \hat{\mathbf{x}}^I$ and $\gamma = 1/\sigma^2$ is the given signal-to-noise ratio (SNR). Let us expand λ_I as $\lambda_I = \sum_{\eta=1}^{N_r} c^2(\eta)d^2(\eta)$, where $c(\eta) = x^I(\eta) - \hat{x}^I(\eta)$. Since the zero elements normally exist in $\mathbf{c}$, we can rewrite λ_I as

$$\lambda_I = \sum_{o=1}^{N_F} c^2(\tau(o))d^2(\tau(o)), \tag{2.29}$$

where N_F denotes the number of non-zero elements of $\mathbf{c}$ and $\{\tau(o)\}_{o=1}^{N_F}$ indicate the indices of non-zero elements in $\mathbf{c}$.

Careful inspection of (2.29) reveals that the phase distribution of the Nakagami-m channel has little impact on the distribution of $d^2(\tau(o))$, which implies that nearly the same system performance will be resulted over Nakagami-m channels with either uniform or non-uniform phases. Moreover, it also reveals that the derivation of PEP in (2.28) requires the probability density function (PDF) of the sum of $d^2(\tau(o))$. However, to the best of our knowledge, this PDF is not available for $m \neq 1$. Hence, for the ease of analysis, we only discuss the case with $m = 1$, in which $d^2(\tau(o)) \sim \text{Gamma}(L, 1)$ with $L = N_t - N_r + 1$ [10]. In [11], this PDF is given by

$$f_{\lambda_I}(y) = \prod_{j=1}^{N_F}\left(\frac{\beta_1}{\beta_j}\right)^L \left[\sum_{k=0}^{+\infty} \frac{\delta_k y^{N_F L+k-1} e^{-y/\beta_1}}{\beta_1^{N_F L+k}\Gamma(N_F L+k)}\right] H_0(x), \tag{2.30}$$

where $H_0(x)$ and $\Gamma(\cdot)$ denote the Heaviside step function and gamma function, respectively, and $\{\beta_j\}_{j=1}^{N_F}$ are the eigenvalues of $\mathbf{A} = \mathbf{BC}$ with $\beta_1 = \min\{\beta_j\}_{j=1}^{N_F}$. Here, $\mathbf{B}$ is an $N_F \times N_F$ diagonal matrix with the diagonal elements $\{c^2(\tau(o))\}_{o=1}^{N_F}$ and $\mathbf{C}$ is an $N_F \times N_F$ matrix

$$\mathbf{C} = \begin{bmatrix} 1 & \sqrt{\rho_I} & \cdots & \sqrt{\rho_I} \\ \sqrt{\rho_I} & \ddots & \ddots & \vdots \\ \vdots & \ddots & \ddots & \sqrt{\rho_I} \\ \sqrt{\rho_I} & \cdots & \sqrt{\rho_I} & 1 \end{bmatrix}, \tag{2.31}$$

where $\rho_I = E[(d^2(\tau(o)) - L)(d^2(\tau(\hat{o})) - L)]/L$ denotes the Pearson product-moment correlation coefficient for any two different Gamma random variables (RVs) $d^2(\tau(o))$ and $d^2(\tau(\hat{o}))$ in $\mathbf{D}^2$. The parameter δ_k is given by

$$\begin{cases} \delta_0 &= 1, \\ \delta_{k+1} &= \frac{L}{k+1}\sum\limits_{l=1}^{k+1}\sum\limits_{j=1}^{N_F}(1-\frac{\beta_1}{\beta_j})^l\delta_{k+1-l}, \quad k = 0, 1, \ldots \end{cases}.$$

By averaging (2.28) with (2.30), we have

$$\Pr(\mathbf{x}_I \to \hat{\mathbf{x}}_I) = E_{\lambda_I}\left[Q\left(\sqrt{\frac{\lambda_I}{2}\gamma}\right)\right] \leq E_{\lambda_I}\left[\frac{1}{2}e^{-\lambda_I\gamma/4}\right], \tag{2.32}$$

where the Chernoff bound $Q(x) \leq \frac{1}{2}e^{-x^2/2}$ is applied. Using the integration formula [12, Eq. (3.381, 4)], (2.32) can be calculated as

$$\begin{aligned}\Pr(\mathbf{x}_I \to \hat{\mathbf{x}}_I) &\leq \frac{1}{2}\prod_{j=1}^{N_F}\left(\frac{\beta_1}{\beta_j}\right)^L \sum_{k=0}^{+\infty}\frac{\delta_k \beta_1^{-N_F L-k}}{\Gamma(N_F L+k)}\int_0^{+\infty} y^{N_F L+k-1}e^{-(\frac{\gamma}{4}+\frac{1}{\beta_1})y}dy \\ &= \frac{1}{2}\prod_{j=1}^{N_F}\left(\frac{\beta_1}{\beta_j}\right)^L\left(\frac{\beta_1\gamma}{4}+1\right)^{-N_F L}\sum_{k=0}^{+\infty}\frac{\delta_k}{(\beta_1\gamma/4+1)^k}.\end{aligned} \tag{2.33}$$

At high SNR ($\gamma \to +\infty$), the PEP can be simplified as

$$\Pr(\mathbf{x}_I \to \hat{\mathbf{x}}_I) \approx \frac{\prod_{j=1}^{N_F}\left(\frac{\beta_1}{\beta_j}\right)^L}{2}\left(\frac{\beta_1\gamma}{4}+1\right)^{-N_F L}, \tag{2.34}$$

where (2.34) is obtained by $\sum_{k=0}^{+\infty}\frac{\delta_k}{(\beta_1\gamma/4+1)^k} = 1$.

With the PEP, the upper bound on ABEP P_E^I for the n_1 bits now can be evaluated by (2.27). It can be observed that a high SNR approximation of P_E^I is the linear combination of PEP, which is only dominated by the smallest exponent of γ with $N_F = 1$. Therefore, at very high SNR we have

$$P_E^I \approx \frac{1}{2^{n_1+1}n_1}\sum_{\mathbf{x}^I}\sum_{\hat{\mathbf{x}}^I \neq \mathbf{x}^I} N(\mathbf{x}^I, \hat{\mathbf{x}}^I)\left(\frac{\beta_1\gamma}{4}+1\right)^{-L}. \tag{2.35}$$

In the same manner, an upper bound on ABEP P_E^Q for the n_2 bits can be obtained. Finally, an upper bound on ABEP for GPQSM can be expressed as

$$P_E = \frac{n_1 P_E^I + n_2 P_E^Q}{n_1 + n_2}, \tag{2.36}$$

where it reveals that the diversity order of our scheme is L. We see that the diversity order depends on the system configuration only and is independent of N_p and M.

2.2.3 Performance Evaluation

In this subsection, we conduct computer simulations to examine the ABEP performance of GPQSM and make comparison with that of the conventional generalised pre-coding aided spatial modulation (GPSM) and MIMO schemes.

In Fig. 2.5, the ABEP performances of various GPQSM schemes under spectral efficiencies of 8 and 32 bps/Hz are drawn with $m = 1$. In order to match 8 bps/Hz for 8×4 MIMO setup, 4-QAM is chosen for GPQSM with $N_p = 2$ and the MIMO scheme while 16-QAM is chosen for GPQSM with $N_p = 1$. On the other hand, to match 32 bps/Hz for 10×8 MIMO setup, 16-QAM is adopted for both the GPQSM with $N_p = 6$ and MIMO schemes. It can be seen that GPQSM schemes obtain an overwhelming performance gain over the MIMO schemes in the high SNR regime. Note that the performance of the MIMO scheme becomes better than the GPQSM scheme with $N_p = 1$ for the 8×4 system. This is because using 16-QAM for GPQSM leads to a worse performance than using 4-QAM for the MIMO scheme, which cannot be compensated for by the performance gain from the spatial domain.

Figure 2.6 investigates the ABEP performance of a 8×4 GPQSM scheme using 4-QAM modulation and a 8×4 GPSM scheme using 8-QAM modulation both with $N_p = 2$ and $m = 2$. The performance of the proposed suboptimal detector is also examined. At ABEP $= 10^{-5}$, it can be seen that the proposed GPQSM scheme achieves about a 3.5 dB SNR gain with respect to the GPSM scheme under the

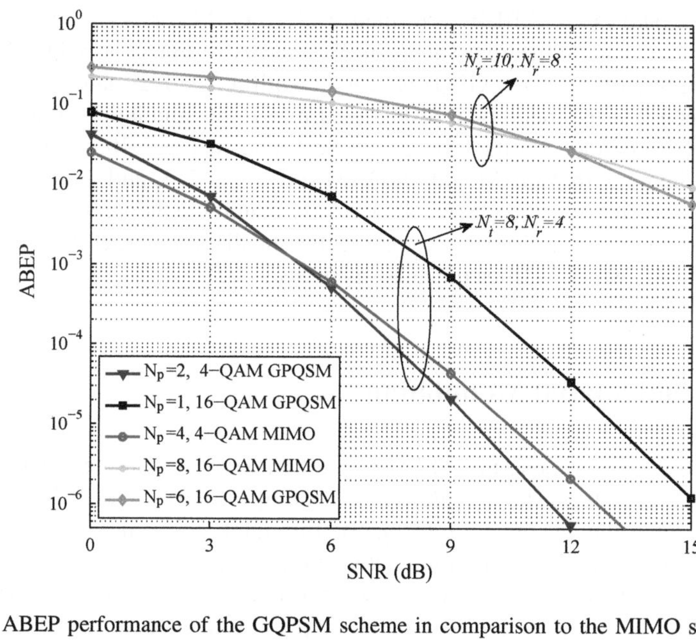

Fig. 2.5 ABEP performance of the GQPSM scheme in comparison to the MIMO scheme with $m = 1$

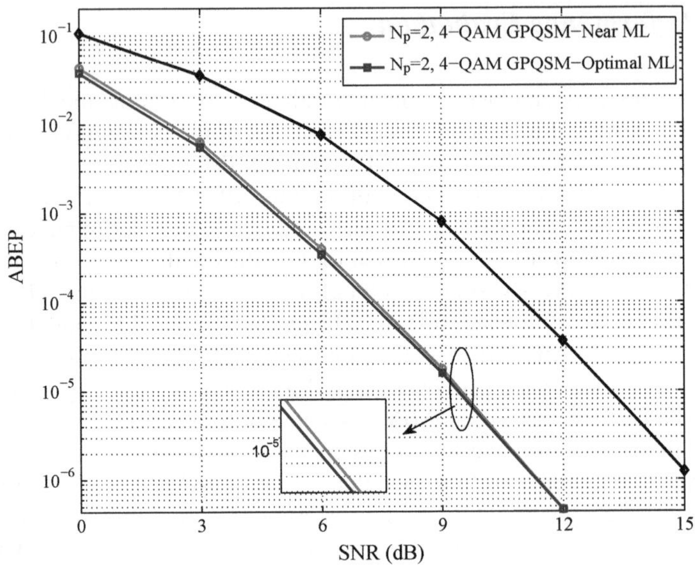

Fig. 2.6 ABEP performance of GPQSM and GPSM with $N_t = 8$ and $N_r = 4$ for $m = 2$

spectral efficiency of 8 bps/Hz. From Fig. 2.6, we can see that the proposed detector achieves nearly the same performance as the ML detector.

Figure 2.7 shows the ABEP performance of GPQSM for $N_t = \{8, 12\}$, $N_r = \{4, 8\}$ and $m = 1$ with different values of N_p and M. In Fig. 2.7a, 4-QAM and 16-QAM are applied for a 8×4 GPQSM scheme with $N_p = 1, 2$, and 3, respectively. It can be seen that the theoretical results well match the simulation counterparts in the high SNR regime. Similar results can be found in Fig. 2.7b, where a 12×8 GPQSM scheme with 4-QAM and 16-QAM for $N_p = 2, 3$, and 4 are considered. The simulation curves of GPQSM also perfectly match the theoretical curves in the high SNR regime. From Fig. 2.7, we can see that the diversity order does not change with N_p and M, which verifies the analysis in Sect. 2.2.2.

2.3 Virtual Spatial Modulation

In this section, we propose the VSM scheme that operates on the virtual parallel channels resulting from the singular value decomposition (SVD) of MIMO channels. The VSM scheme conveys information through both the indices of the virtual parallel channels and the M-ary modulated symbols, applying to any configuration of transceiver antennas. We derive a closed-form upper bound on the ABEP of the VSM scheme, which takes into account the impact of imperfect channel estimation. Moreover, the asymptotic ABEP is also studied, which characterizes the ABEP error

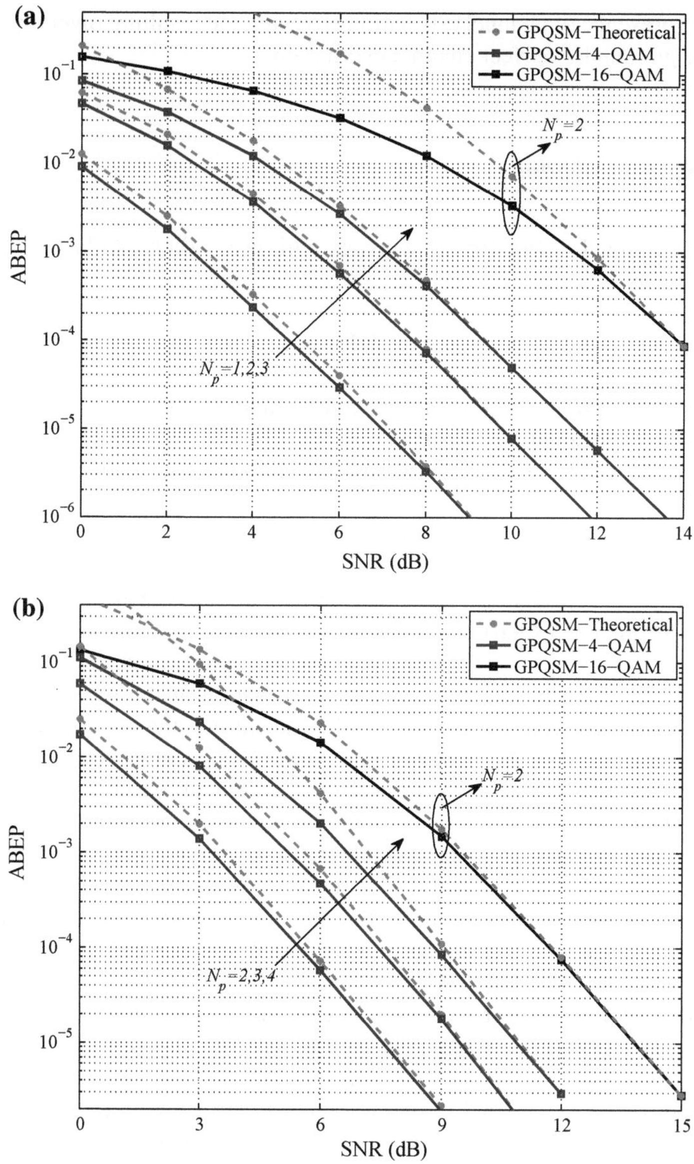

Fig. 2.7 ABEP performance of GPQSM with $m = 1$ for different configurations. **a** $N_t = 8$ and $N_r = 4$. **b** $N_t = 12$ and $N_r = 8$

floor in the presence of channel estimation error and the resulting diversity order and the coding gain under perfect channel estimation.

2.3.1 System Model

The received signal vector for all schemes with pre-coding can be expressed in a general form as

$$\mathbf{y} = \mathbf{HPx} + \mathbf{w}, \tag{2.37}$$

where $\mathbf{x}$ is the signal vector, $\mathbf{P}$ is the pre-coding matrix and $\mathbf{w} \in \mathbb{C}^{N_r \times 1}$ is an AWGN vector with zero mean and covariance matrix $N_0\mathbf{I}_{N_r}$.

The well-known parallel decomposition of $\mathbf{H}$ using SVD is given by

$$\mathbf{H} = \mathbf{U\Sigma V}^H, \tag{2.38}$$

where $\mathbf{U} \in \mathbb{C}^{N_r \times N_r}$ and $\mathbf{V} \in \mathbb{C}^{N_t \times N_t}$ are unitary matrices, and $\mathbf{\Sigma} \in \mathbb{C}^{N_r \times N_t}$ is the singular value matrix, which contains singular values $\{\sigma_i\}_{i=1}^m$ with $m = \min\{N_t, N_r\}$. Inspired by SVD, we aim to transmit the spatial bits through the virtual parallel channels in $\mathbf{\Sigma}$. If we have perfect channel estimation at the receiver and perfect feedback of the CSI from the receiver to the transmitter, the pre-coding matrix can be simply set to be $\mathbf{P} = \mathbf{V}$. However, the perfect CSI is always unavailable at the receiver in practical systems. Therefore, to be practical we take into account the imperfect channel estimation. The estimated channel with errors at the receiver can be modeled as $\tilde{\mathbf{H}} = \mathbf{H} + \mathbf{H}_e$, where $\mathbf{H}_e$ denotes the channel estimation error matrix whose entries are independent and identically distributed complex Gaussian RVs with zero mean and variance δ_e^2. Note that $\mathbf{H}_e$ is independent of $\mathbf{H}$ and $\tilde{\mathbf{H}}$ depends on $\mathbf{H}$ with the correlation coefficient $\rho = 1/(1+\delta_e^2)$. It can be seen that the perfect channel estimation is the special case of $\delta_e^2 = 0$ $(\rho = 1)$. In Sect. 2.3.3, we will consider two situations for determining ρ: (1) fixed ρ: the value of estimation error is fixed for all SNRs; (2) unfixed ρ: the estimation error changes with the given SNR as $\delta_e^2 = 1/(L\gamma)$, where $\gamma = 1/N_0$ denotes the SNR and L denotes the number of pilot symbols used for channel estimation [13]. Note that the feedback error can be incorporated into $\mathbf{H}_e$. For ease of exposition, we simply consider perfect feedback.

Due to channel estimation, the receiver knows $\tilde{\mathbf{H}}$ instead of $\mathbf{H}$. Performing SVD to $\tilde{\mathbf{H}}$, we have

$$\tilde{\mathbf{H}} = \tilde{\mathbf{U}}\tilde{\mathbf{\Sigma}}\tilde{\mathbf{V}}^H, \tag{2.39}$$

where $\tilde{\mathbf{U}} \in \mathbb{C}^{N_r \times N_r}$ and $\tilde{\mathbf{V}} \in \mathbb{C}^{N_t \times N_t}$ are the unitary matrices, and $\tilde{\mathbf{\Sigma}} \in \mathbb{C}^{N_r \times N_t}$ is the singular value matrix. The right unitary matrix $\tilde{\mathbf{V}}$ is fed back to the transmitter for pre-coding purposes. Under a rich scattering environment assumption, $\tilde{\mathbf{\Sigma}} \in \mathbb{C}^{N_r \times N_t}$

contains non-zero singular values $\{\tilde{\sigma}_i\}_{i=1}^{m}$, which can also provide m virtual parallel channels for carrying spatial bits. To perform index modulation on the m virtual parallel channels, however, the signal vector $\mathbf{x}$ has to be carefully designed to match the pre-coding matrix $\tilde{\mathbf{V}}$. To this end, we have to classify the SVD of $\tilde{\mathbf{H}}$ into two situations. When $N_t \leq N_r$, the SVD of $\tilde{\mathbf{H}}$ can be rewritten as

$$\tilde{\mathbf{H}} = [\tilde{\mathbf{U}}_1, \tilde{\mathbf{U}}_0][\tilde{\boldsymbol{\Sigma}}_1, \mathbf{0}]^T \tilde{\mathbf{V}}^H, \tag{2.40}$$

where $\tilde{\mathbf{U}}_1 \in \mathbb{C}^{N_r \times N_t}$ and $\tilde{\mathbf{U}}_0 \in \mathbb{C}^{N_r \times (N_r - N_t)}$ are the sub-matrices of $\tilde{\mathbf{U}}$, which collect the left-singular vectors corresponding to nonzero and zero singular values, respectively, and $\tilde{\boldsymbol{\Sigma}}_1 \in \mathbb{C}^{N_t \times N_t}$ is the non-zero singular value matrix. In this case, the signal vector (a.k.a., the information vector) is set to be $\mathbf{x} \in \mathbb{C}^{N_t \times 1}$. On the other hand, when $N_t > N_r$, the SVD of $\tilde{\mathbf{H}}$ can be rewritten as

$$\tilde{\mathbf{H}} = \tilde{\mathbf{U}}[\tilde{\boldsymbol{\Sigma}}_1, \mathbf{0}][\tilde{\mathbf{V}}_1, \tilde{\mathbf{V}}_0]^H, \tag{2.41}$$

where $\tilde{\mathbf{V}}_1 \in \mathbb{C}^{N_t \times N_r}$ and $\tilde{\mathbf{V}}_0 \in \mathbb{C}^{N_t \times (N_t - N_r)}$ are sub-matrices of $\tilde{\mathbf{V}}$, which collect the right-singular vectors corresponding to nonzero and zero singular values, respectively. In this case, the signal vector is set to be $\tilde{\mathbf{x}} \in \mathbb{C}^{N_t \times 1} = [\mathbf{x}^T, \mathbf{z}_0^T]^T$, where $\mathbf{x} \in \mathbb{C}^{N_r \times 1}$ is the information vector corresponding to the m virtual parallel channels and $\mathbf{z}_0 \in \mathbb{C}^{(N_t - N_r) \times 1}$ denotes the all-zero vector.

VSM Transmitter

The transmitter structure of VSM is depicted in Fig. 2.8. In VSM, N_p out of m virtual parallel channels are activated for transmitting the real (imaginary) parts of the modulated symbols. In order to modulate the spatial bits, only $N = 2^{\lfloor \log_2 (C(m, N_p) \rfloor}$ combinations are selected as the legitimate antenna combinations. Note that the number of virtual parallel channels is m, which equals the minimum between the numbers of transmit and receive antennas. Therefore, the VSM scheme contains both benefits of transmitter and receiver PSM schemes.

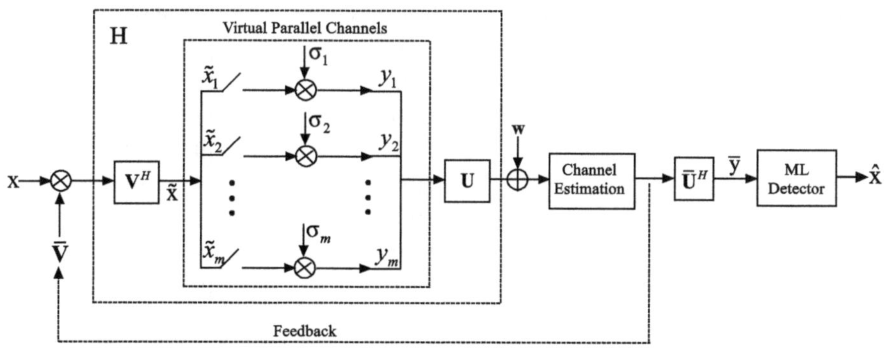

Fig. 2.8 The schematic of VSM

Specifically, VSM works as follows. At each time slot, $d = N_p \log_2(M) + 2\lfloor \log_2(\mathrm{C}(m, N_p)) \rfloor$ information bits are fed into the transmitter and divided into three parts. The first part of $N_p \log_2(M)$ bits are mapped into a symbol vector $\mathbf{s} = [s_1, \ldots, s_{N_p}]^T$, where $s_\varphi = s_\varphi^I + js_\varphi^Q, \varphi \in \{1, \ldots, N_p\}$, with s_φ^I and s_φ^Q denoting the in-phase and quadrature parts of s_φ, respectively. Note that $s_i^I \in \mathcal{X}^I$ and $s_i^Q \in \mathcal{X}^Q$, where $\mathcal{X}^I$ and $\mathcal{X}^Q$ denote the in-phase and quadrature sets of $\mathcal{X}$, respectively. The second part of $\lfloor \log_2(\mathrm{C}(m, N_p)) \rfloor$ bits select an active channel combination $I_v^I = \{i_v^I(1), \ldots, i_v^I(N_p)\}$, where $i_v^I(\alpha) \in \{1, \ldots m\}$ with $v \in \{1, \ldots, N\}$ and $\alpha \in \{1, \ldots, N_p\}$, for transmitting the real part of $\mathbf{s}$, yielding $\mathbf{x}^I = [\ldots, s_1^I, \ldots, s_{N_p}^I, \ldots]$. Similarly, the third part of $\lfloor \log_2(\mathrm{C}(m, N_p)) \rfloor$ bits select an active channel combination $I_\mu^Q = \{i_\mu^Q(1), \ldots, i_\mu^Q(N_p)\}$, where $i_\mu^Q(\varsigma) \in \{1, \ldots m\}$ with $\mu \in \{1, \ldots, N\}$ and $\varsigma \in \{1, \ldots, N_p\}$, for transmitting the imaginary part of $\mathbf{s}$, yielding $\mathbf{x}^Q = [\ldots, s_1^Q, \ldots, s_{N_p}^Q, \ldots]$. Finally, the information vector is generated by combining the in-phase and quadrature parts:

$$\mathbf{x} = \mathbf{x}^I + j\mathbf{x}^Q. \tag{2.42}$$

As an example, we present a mapping table of VSM for $N_t = 4$, $N_r = 2$, $N_p = 1$, and 4-QAM in Table 2.1. In this example, we have $d = 4$. Suppose that the information bits $q = [0\ 1\ 1\ 0]$ are transmitted at a given time slot. The first $N_p \log_2(M) = 2$ bits $[0\ 1]$ are mapped into $\mathbf{s} = [-1 - j]$. Then, $\mathbf{s}$ is divided into in-phase and quadrature parts $\mathbf{s}^I = [-1]$ and $\mathbf{s}^Q = [-1]$, respectively. The second $\lfloor \log_2(\mathrm{C}(N_r, N_p)) \rfloor = 1$ bit [1] indicates the active channel combination $I_2^I = \{2\}$ to transmit $\mathbf{s}^I$, resulting in the vector $\mathbf{x}^I = [0, -1]^T$. The last $\lfloor \log_2(\mathrm{C}(N_r, N_p)) \rfloor = 1$ bit [0] indicates the active channel combination $I_1^Q = \{1\}$ to transmit $\mathbf{s}^Q$ resulting in the vector $\mathbf{x}^Q = [1, 0]^T$. Finally, the information vector is obtained as $\mathbf{x} = [-j, -1]^T$.

Before transmitted into the wireless channel, the information vector is pre-coded by $\mathbf{P} = \bar{\mathbf{V}}$, which represents $\tilde{\mathbf{V}}$ for $N_t \leq N_r$ and $\tilde{\mathbf{V}}_1$ for $N_t > N_r$, respectively.

Table 2.1 Mapping table for VSM with $N_t = 4$, $N_r = 2$, $N_p = 1$ and 4-QAM

Input bits	Signal vector	Input bits	Signal vector
0000	$[-1 + j, 0]^T$	1000	$[1 + j, 0]^T$
0001	$[-1, j]^T$	1001	$[1, j]^T$
0010	$[j, -1]^T$	1010	$[j, 1]^T$
0011	$[0, -1 + j]^T$	1011	$[0, 1 + j]^T$
0100	$[-1 - j, 0]^T$	1100	$[1 - j, 0]^T$
0101	$[-1, -j]^T$	1101	$[1, -j]^T$
0110	$[-j, -1]^T$	1110	$[-j, 1]^T$
0111	$[0, -1 - j]^T$	1111	$[0, 1 - j]^T$

VSM Receiver

The receiver structure of VSM is depicted in Fig. 2.8. The received signal vector after receiver shaping can be expressed in a general form as

$$\bar{\mathbf{y}} = \bar{\mathbf{U}}^H \mathbf{y} = \bar{\mathbf{U}}^H \mathbf{H} \bar{\mathbf{V}} \mathbf{x} + \bar{\mathbf{U}}^H \mathbf{w}, \tag{2.43}$$

where $\bar{\mathbf{U}}$ represents $\tilde{\mathbf{U}}_1$ or $\tilde{\mathbf{U}}$ for $N_t \leq N_r$ or $N_t > N_r$, respectively.

Based on (2.43), we propose two detection methods corresponding to whether or not the receiver has the knowledge of the correlation coefficient ρ. When ρ is known to the receiver, the optimal ML detector can be derived as

$$\hat{\mathbf{x}} = \arg\min_{\mathbf{x}} \|\bar{\mathbf{y}} - \rho \tilde{\mathbf{\Sigma}}_1 \mathbf{x}\|^2. \tag{2.44}$$

The reason why (2.44) is optimal will be clarified later. On the other hand, when ρ is unknown to the receiver, it is reasonable to assume perfect channel estimation for detection, under which the detector follows from (2.44) by setting $\rho = 1$:

$$\hat{\mathbf{x}} = \arg\min_{\mathbf{x}} \|\bar{\mathbf{y}} - \tilde{\mathbf{\Sigma}}_1 \mathbf{x}\|^2. \tag{2.45}$$

However, both detectors in (2.44) and (2.45) impose great computational burden to the receiver when N_r or M goes to a very large value. To solve this problem, we can resort to the idea proposed in Sect. 2.2.1.

2.3.2 Average Bit Error Probability Analysis

We derive a closed-form upper bound on the ABEP of VSM that employs the detector in (2.44). To study the diversity order and coding gain, we further derive the asymptotic ABEP of VSM. We note that it may be difficult to obtain closed-form upper bounded and asymptotic ABEPs of VSM that adopts the detector in (2.45). Fortunately, as will be verified in Sect. 2.3.3 that both detectors achieve nearly the same ABEP performance, the analysis for the detector in (2.44) suffices.

Analysis of ABEP

Based on the first-order autoregressive model, we can express the true channel with the estimated one as $\mathbf{H} = \rho\tilde{\mathbf{H}} + \mathbf{G}$, where $\mathbf{G} \in \mathbb{C}^{N_r \times N_t}$ is a complex Gaussian matrix independent of $\tilde{\mathbf{H}}$, whose entries have zero mean and variance $1 - \rho$. Accordingly, $\bar{\mathbf{y}}$ can be rewritten from (2.43) as

$$\bar{\mathbf{y}} = \rho \tilde{\mathbf{\Sigma}}_1 \mathbf{x} + \mathbf{q}, \tag{2.46}$$

where $\mathbf{q} = \bar{\mathbf{U}}^H \mathbf{G} \bar{\mathbf{V}} \mathbf{x} + \bar{\mathbf{U}}^H \mathbf{w}$. Since multiplying a matrix by any unitary matrix does not change the probability distribution, we can conclude that $\bar{\mathbf{U}}^H \mathbf{G} \bar{\mathbf{V}}$ and $\bar{\mathbf{U}}^H \mathbf{w}$ have the same distribution as $\mathbf{G}$ and $\mathbf{w}$, respectively. In addition, since $\mathbf{G}$ and $\mathbf{w}$ are mutually independent, each entry of $\mathbf{q}$ follows the complex Gaussian distribution with zero mean and variance

$$\psi = N_0 + (1-\rho)||\mathbf{x}||^2. \tag{2.47}$$

At this point, it is straightforward to arrive at (2.44) from (2.46), which is optimal for detection.

Denote $\mathbf{c} \triangleq [c_1, \ldots, c_m]^T = \mathbf{x} - \hat{\mathbf{x}}$. In light of (2.44) and (2.46), the conditional PEP of detecting $\hat{\mathbf{x}}$ when $\mathbf{x}$ is transmitted on $\tilde{\mathbf{\Sigma}}_1$ can be calculated as

$$\begin{aligned}
\Pr(\mathbf{x} \to \hat{\mathbf{x}}|\tilde{\mathbf{\Sigma}}_1) &= \Pr(||\bar{\mathbf{y}} - \rho\tilde{\mathbf{\Sigma}}_1\mathbf{x}||^2 > ||\bar{\mathbf{y}} - \rho\tilde{\mathbf{\Sigma}}_1\hat{\mathbf{x}}||^2) \\
&= Q\left(\sqrt{\frac{||\rho\tilde{\mathbf{\Sigma}}_1(\mathbf{x}-\hat{\mathbf{x}})||^2}{2\psi}}\right) \\
&= Q\left(\sqrt{\frac{\sum_{i=1}^m \lambda_i |c_i|^2}{2\psi}}\right),
\end{aligned} \tag{2.48}$$

where $\lambda_i = \rho^2\tilde{\sigma}_i^2, i \in \{1, \ldots, m\}$. From (2.48), it can be seen that the derivation of PEP needs the joint PDF of $\mathbf{\Lambda} = [\lambda_1, \ldots, \lambda_m]^T$, which is given by [14]

$$\begin{aligned}
f(\mathbf{\Lambda}) &= \frac{1}{K_{m,n}} e^{-\sum\limits_{i=1}^{m}\lambda_i} \underbrace{\prod_{i=1}^{m} \lambda_i^{n-m} \prod_{j=i+1}^{m} (\lambda_i - \lambda_j)^2}_{\Upsilon} \\
&= \frac{1}{K_{m,n}} \sum_{\kappa=1}^{L} a_\kappa e^{-\sum\limits_{i=1}^{m}\lambda_i} \lambda_1^{b_\kappa^1} \cdots \lambda_m^{b_\kappa^m},
\end{aligned} \tag{2.49}$$

where $n = \max\{N_r, N_t\}$, $K_{m,n}$ is a normalizing coefficient, L denotes the number of monomials in Υ, a_κ denotes the coefficient of the κ-th monomial in Υ with $\kappa \in \{1, \ldots L\}$, and $\{b_\kappa^i\}_{i=1}^m$ denote the power coefficients of $\{\lambda_i\}_{i=1}^m$ in the κ-th monomial in Υ. An example for all values of $(K_{m,n}, a_\kappa, b_\kappa^1, \ldots, b_\kappa^m)$ with $N_t = 4$ and $N_r = 3$ is given in Table 2.2. By averaging (2.48) with (2.49), we have

$$\begin{aligned}
\Pr(\mathbf{x} \to \hat{\mathbf{x}}) =& E_{\mathbf{\Lambda}}\left[Q\left(\sqrt{\frac{\sum_{i=1}^m \lambda_i |c_i|^2}{2\psi}}\right)\right] \\
\cong& E_{\mathbf{\Lambda}}\left[\frac{1}{12} e^{-\sum\limits_{i=1}^{m}\frac{\lambda_i|c_i|^2}{4\psi}} + \frac{1}{4} e^{-\sum\limits_{i=1}^{m}\frac{\lambda_i|c_i|^2}{3\psi}}\right] && (2.50) \\
=& \frac{1}{12}\int e^{-\sum\limits_{i=1}^{m}\frac{\lambda_i|c_i|^2}{4\psi}} f(\mathbf{\Lambda})d\mathbf{\Lambda} + \frac{1}{4}\int e^{-\sum\limits_{i=1}^{m}\frac{\lambda_i|c_i|^2}{3\psi}} f(\mathbf{\Lambda})d\mathbf{\Lambda}, && (2.51)
\end{aligned}$$

where (2.50) is obtained by [15]

$$Q(x) \cong \frac{1}{12}e^{-\frac{x^2}{2}} + \frac{1}{4}e^{-\frac{2x^2}{3}}. \tag{2.52}$$

According to [12, Eq. (8.312.2)], the first integral of (2.51) can be calculated as

$$\begin{aligned}
&\frac{1}{12}\int e^{\frac{-\sum_{i=1}^{m}\lambda_i|c_i|^2}{4\psi}} f(\boldsymbol{\Lambda})d\boldsymbol{\Lambda} \\
&= \frac{1}{12K_{m,n}}\sum_{\kappa=1}^{L} a_\kappa \int e^{-\sum_{i=1}^{m}(\frac{|c_i|^2}{4\psi}+1)\lambda_i} \lambda_1^{b_\kappa^1}\cdots\lambda_m^{b_\kappa^m} d\boldsymbol{\Lambda} \\
&= \frac{1}{12K_{m,n}}\sum_{\kappa=1}^{L} a_\kappa \int e^{-(\frac{|c_1|^2}{4\psi}+1)\lambda_1}\lambda_1^{b_\kappa^1} d\lambda_1 \cdots \int e^{-(\frac{|c_m|^2}{4\psi}+1)\lambda_m}\lambda_m^{b_\kappa^m} d\lambda_m \\
&= \frac{1}{12K_{m,n}}\sum_{\kappa=1}^{L} a_\kappa \prod_{i=1}^{m}\left(\frac{|c_i|^2}{4\psi}+1\right)^{-(b_\kappa^i+1)} \Gamma(b_\kappa^i+1).
\end{aligned} \tag{2.53}$$

In the same manner, the second integral of (2.51) is given by

$$\frac{1}{4K_{m,n}}\sum_{\kappa=1}^{L} a_\kappa \prod_{i=1}^{m}\left(\frac{|c_i|^2}{3\psi}+1\right)^{-(b_\kappa^i+1)} \Gamma(b_\kappa^i+1). \tag{2.54}$$

By substituting (2.53) and (2.54) into (2.51), the unconditional PEP can be expressed as

$$\begin{aligned}
\Pr(\mathbf{x}\rightarrow\hat{\mathbf{x}}) =& \frac{1}{4K_{m,n}}\sum_{\kappa=1}^{L} a_\kappa \left(\prod_{i=1}^{m}\Gamma(b_\kappa^i+1)\right) \\
&\times\left[\frac{1}{3}\prod_{i=1}^{m}\left(\frac{|c_i|^2}{4\psi}+1\right)^{-(b_\kappa^i+1)} + \prod_{i=1}^{m}\left(\frac{|c_i|^2}{3\psi}+1\right)^{-(b_\kappa^i+1)}\right].
\end{aligned} \tag{2.55}$$

After obtaining the PEP, an upper bound on the ABEP can be readily derived according to the union bounding technique as

$$P_e \le \frac{1}{d2^d}\sum_{\mathbf{x}}\sum_{\hat{\mathbf{x}}\neq\mathbf{x}} N(\mathbf{x},\hat{\mathbf{x}})\Pr(\mathbf{x}\rightarrow\hat{\mathbf{x}}). \tag{2.56}$$

Asymptotic ABEP Analysis

Recall $\psi = N_0 + (1-\rho)||\mathbf{x}||^2$. If we have channel estimation error ($\rho \neq 1$), $N_0 = 1/\gamma$ tends to be zero and $\psi \rightarrow \psi^* = (1-\rho)||\mathbf{x}||^2$ as $\gamma \rightarrow +\infty$. The asymptotic ABEP can be thus expressed from (2.56) as

$$P_e \to \frac{1}{d2^{d+2}K_{m,n}} \sum_{\mathbf{x}} \sum_{\hat{\mathbf{x}}} N(\mathbf{x}, \hat{\mathbf{x}}) \sum_{\kappa=1}^{L} a_\kappa \left(\prod_{i=1}^{m} \Gamma(b_\kappa^i + 1) \right)$$
$$\times \left[\frac{1}{3} \prod_{i=1}^{m} \left(\frac{|c_i|^2}{4\psi^*} + 1 \right)^{-(b_\kappa^i+1)} + \prod_{i=1}^{m} \left(\frac{|c_i|^2}{3\psi^*} + 1 \right)^{-(b_\kappa^i+1)} \right]. \tag{2.57}$$

The above equation reveals that the ABEP tends to be a constant irrelevant to the SNR in the high SNR region, which, as reflected in the ABEP performance, creates an error floor.

On the other hand, if no channel estimation error exists ($\rho = 1$), we have $\psi = 1/\gamma$ and the asymptotic ABEP can be derived as

$$P_e \to \frac{1}{d2^{d+2}K_{m,n}} \sum_{\mathbf{x}} \sum_{\hat{\mathbf{x}}} N(\mathbf{x}, \hat{\mathbf{x}}) \sum_{\kappa=1}^{L} a_\kappa \left(\prod_{i=1}^{m} \Gamma(b_\kappa^i + 1) \right)$$
$$\times \left[\frac{1}{3} \prod_{i=1}^{m} \left(\frac{|c_i|^2\gamma}{4} \right)^{-(b_\kappa^i+1)} + \prod_{i=1}^{m} \left(\frac{|c_i|^2\gamma}{3} \right)^{-(b_\kappa^i+1)} \right]. \tag{2.58}$$

From (2.58), we see that the high SNR approximation of P_e turns out to be a linear combination of PEPs, which are dominated by the sum of the terms associated with the largest exponent (equivalently the smallest value among $\{b_\kappa^i\}_{i=1}^m$ for all κ(s)) of γ, For instance, from Table 2.2 with $N_t = 4$ and $N_r = 3$, the smallest coefficient is 1. From the above analysis, the asymptotic ABEP in (2.58) can be further approximated as

$$P_e \to \frac{\gamma^{-d_{min}}}{d2^{d+2}K_{m,n}} \sum_{\kappa, d_{min}} \sum_{s,\hat{s}\in\mathcal{X}} a_\kappa \left(\prod_{i=1}^{m} \Gamma(b_\kappa^i + 1) \right)$$
$$\times\ N(\mathbf{x}, \hat{\mathbf{x}}) \left(\frac{|s-\hat{s}|^{-2d_{min}}}{3 \cdot 4^{-d_{min}}} + \frac{|s-\hat{s}|^{-2d_{min}}}{3^{-d_{min}}} \right), \tag{2.59}$$

Table 2.2 A look-up table for $N_t = 4$, $N_r = 3$ ($K_{3,4} = 144$)

κ	a_κ	b_κ^1	b_κ^2	b_κ^3	κ	a_κ	b_κ^1	b_κ^2	b_κ^3	κ	a_κ	b_κ^1	b_κ^2	b_κ^3
1	1	1	3	5	8	2	3	4	2	15	1	3	1	5
2	−2	1	4	4	9	1	3	5	1	16	−2	4	1	4
3	1	1	5	3	10	−2	2	2	5	17	1	5	1	3
4	2	2	3	4	11	2	3	2	4	18	−2	5	2	2
5	2	2	4	3	12	2	4	2	3	19	1	5	3	1
6	−2	2	5	2	13	2	4	3	2					
7	−6	3	3	3	14	−2	4	4	1					

where $d_{min} = n - m + 1$ is the diversity order achieved by the VSM scheme and $\sum_{\kappa, d_{min}}$ denotes the sum applying to all κ(s) for which there exists the coefficient d_{min}. From (2.59), we see that similar to the existing pre-coding schemes, the diversity order of VSM is determined by the configuration of the transceiver antennas only and is irreverent to N_p. By definition [16], the coding gain achieved by the VSM scheme can be derived from (2.59) as

$$G_c = \left[\frac{1}{K_{m,n} d 2^{d+2}} \sum_{\kappa, d_{min}} \sum_{s, \hat{s} \in \mathcal{X}} a_\kappa \left(\prod_{i=1}^{m} \Gamma(b_\kappa^i + 1) \right) \right.$$
$$\left. \times \; N(\mathbf{x} \to \hat{\mathbf{x}}) \left(\frac{|s - \hat{s}|^{-2d_{min}}}{3 \cdot 4^{-d_{min}}} + \frac{|s - \hat{s}|^{-2d_{min}}}{3^{-d_{min}}} \right) \right]^{-\frac{1}{d_{min}}}, \tag{2.60}$$

which satisfies

$$P_e \cong (G_c \gamma)^{-d_{min}}, \tag{2.61}$$

in the high SNR region.

2.3.3 *Performance Evaluation*

In this subsection, we conduct computer simulations to evaluate the ABEP performance of VSM. The 4-QAM is assumed for all schemes.

Performance Comparison with Existing Pre-coding Schemes

We evaluate the ABEP performance of VSM for the case of $N_t > N_r$ in Figs. 2.9 and 2.10, where the GPQSM scheme is considered for comparison. The simulation parameters for Fig. 2.9 are chosen as $N_t = 4$, $N_r = 2$, and $N_p = 1$, and those for Fig. 2.10 are $N_t = 8$, $N_r = 4$, and $N_p = 2$. Since the proposed low-complexity detector achieves almost the same performance as the optimal ML detector, we have removed its ABEP curves for figure clarity. To examine the effect of channel estimation, we consider $\rho = 1, 0.95$, and 0.91, which correspond to $\delta_e^2 = 0, 0.01$, and 0.05, respectively. The situation of unfixed ρ will be discussed later. For references, the analytical ABEP upper bounds and asymptotic results of VSM are also added in the figures. It is observed from Figs. 2.9 and 2.10 that, as expected, under imperfect channel estimation, error floors appear, whose levels are accurately characterized by the analytical results. In addition, the error floor increases with the channel estimation error, applying to both schemes. However, the floor level for VSM is much lower. The above observation can be explained by (2.46), which reveals that the power of the equivalent noise $\mathbf{q}$ scales with the channel estimation error instead of the SNR. On the other hand, under perfect channel estimation, VSM achieves a diversity order of $N_t - N_r + 1$ as GPQSM while obtaining a much larger coding gain. For example,

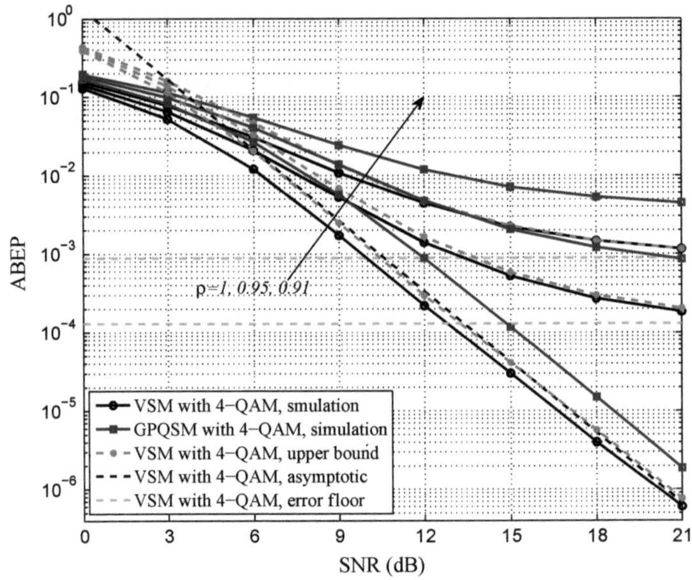

Fig. 2.9 Comparison between VSM and GPQSM in terms of ABEP with $\rho = 1, 0.95, 0.91$ for $N_t = 4$, $N_r = 2$, $N_p = 1$, and 4-QAM

as seen from Fig. 2.9, about 2 dB SNR gain is obtained at an ABEP value of 10^{-5}. This superiority can be accounted for the more diverse channels resulting from the SVD.

The case of $N_t < N_r$ is considered in Fig. 2.11, where $N_t = 2$, $N_r = 4$, $N_p = 1$, and the generalized SM (GSM) scheme is chosen as a comparison representative. For fair comparison, we deprive VSM of the index modulation on the real(imaginary) parts of the received signals, which means VSM activates a single channel to convey a 4-QAM symbol as GSM. It should be noted that as only a single channel is active, GSM degenerates to SM. For figure clarity, only the upper bounded ABEP curve with $\rho = 1$ for VSM is added. From the figure, we see that under perfect channel estimation, GSM outperforms VSM in the low-to-high SNR region. This can be understood since a diversity order of $N_r - N_t + 1$ is achieved by VSM, while a higher diversity order, i.e., N_r, is achieved by GSM. On the other hand, we see that under imperfect channel estimation both VSM and GSM result in error floors, while the level for GSM is much lower. This can be accounted for the SVD used in VSM, which converts MIMO channels into parallel channels by scarifying some copies of transmitted signals. Since besides its worse performance, VSM requires an additional feedback link, VSM is not recommended to be used as a transmitter-side SM scheme in practice.

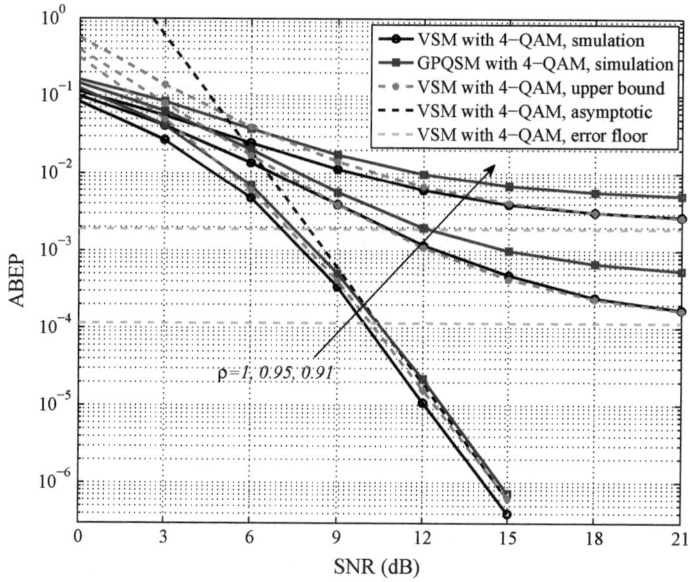

Fig. 2.10 Comparison between VSM and GPQSM in terms of ABEP with $\rho = 1, 0.95, 0.91$ for $N_t = 8$, $N_r = 4$, $N_p = 2$, and 4-QAM

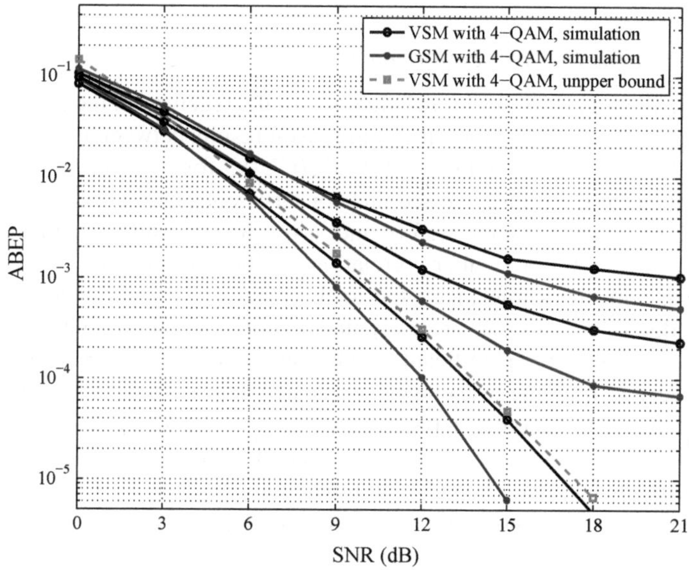

Fig. 2.11 Comparison between VSM and GSM in terms of ABEP with $\rho = 1, 0.95, 0.91$ for $N_t = 2$, $N_r = 4$, $N_p = 1$, and 4-QAM

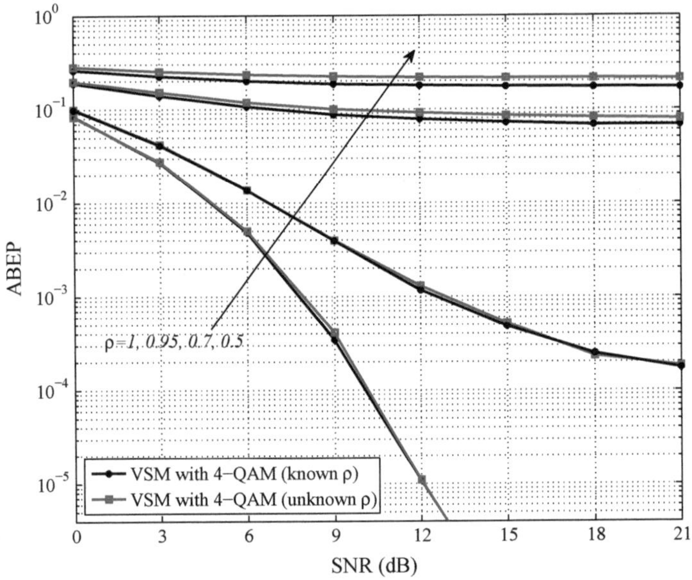

Fig. 2.12 Comparison of the performances of VSM receivers (2.25) and (2.26) in terms of ABEP with $\rho = 1, 0.95, 0.7, 0.5$ for $N_t = 8$, $N_r = 4$, and $N_p = 2$

Impact of ρ *on The Receiver Performance*

We examine the impact of the knowledge of ρ to the receiver on the performance of VSM systems. Figure 2.12 shows the comparison results between the ABEPs of the detectors (2.44) and (2.45) with $\rho = 1, 0.95, 0.7, 0.5$ for $N_t = 8$, $N_t = 4$, $N_p = 2$ and 4QAM. From the figure, we see that for large correlation coefficients such as $\rho = 1$ and 0.95, nearly the same performance is resulted no matter whether ρ is known to the receiver or not. For smaller correlation coefficients such as $\rho = 0.7, 0.5$, the performance with the knowledge of ρ to the receiver is superior to that without the knowledge of ρ. This can be easily understood since the influence of channel estimation becomes more significant as ρ becomes smaller and the knowledge of ρ helps the receiver tract the true channel more accurately. However, we note that as the performance gap becomes apparent, the ABEP value also goes very large, e.g., on an order of 10^{-1} for $\rho = 0.7, 0.5$, which is intolerable for most applications of wireless communications. Therefore, in practice the receiver (2.45) is preferred for VSM systems in consideration of its simplicity in implementation, since usually the true value of ρ is difficult to obtain.

Then, we consider the impact of a different channel estimation model on the ABEP performance of VSM. It is assumed that ρ changes with the SNR according to $\rho = l\gamma/(l\gamma + 1)$, where l is chosen to be 2. Figure 2.13 shows the comparison results between performances of VSM and GPQSM with unfixed ρ for two different system configurations, which are (1) $N_t = 4$, $N_r = 2$, and $N_p = 1$; (2) $N_t = 8$,

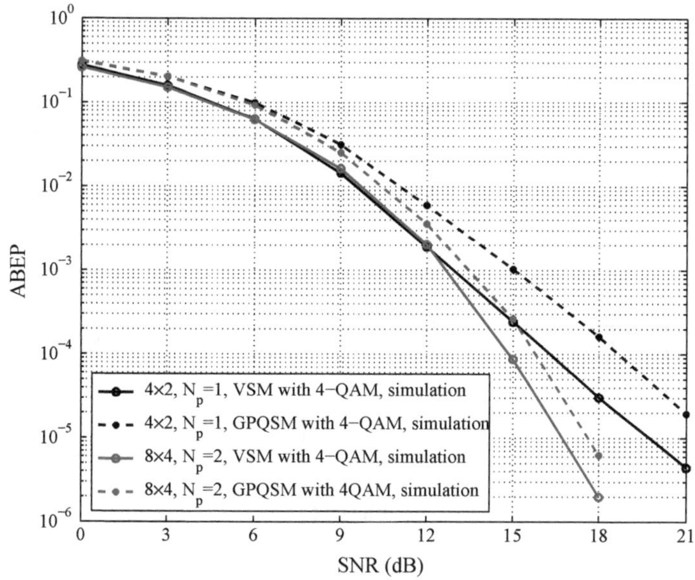

Fig. 2.13 Comparison between VSM and GSM in terms of ABEP with unfixed ρ for two different configurations: (1) $N_t = 4$, $N_r = 2$, and $N_p = 1$; (2) $N_t = 8$, $N_r = 4$, and $N_p = 2$

$N_r = 4$, and $N_p = 2$. From the figure, it can be seen that different from what we observe under the situation of fixed ρ, the error floors never appear for both VSM and GPQSM. This is because ρ approaches one, which happens to be the perfect channel estimation case, in the high SNR region. Therefore, the asymptotic results also apply to the situation of unfixed ρ, which are not shown for figure clarity. On the other hand, from Fig. 2.13, we see that similarly VSM is superior to GPQSM in the whole SNR region.

2.4 Applications to Cooperative Communications

In this section, we demonstrate the potentials and advantages of space domain index modulation techniques in the applications to cooperative communications. We take SSK modulation as a demonstrative example. Specifically, a pragmatic communication strategy for the use of SSK modulation in two-way amplify-and-forward (AF) relaying is proposed. By considering a Nakagami-m fading environment, upper bounded and asymptotic ABEPs are both derived in closed form. The optimal power allocation problem in the minimization of the ABEP is also addressed.

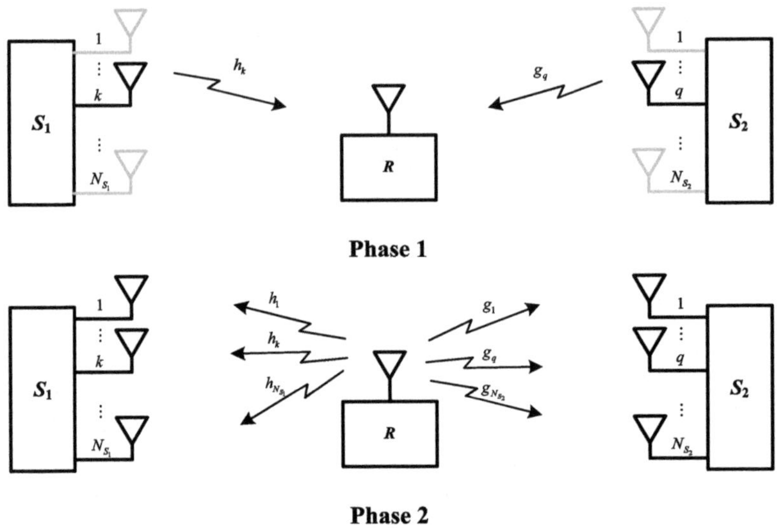

Fig. 2.14 The schematic of the two-way relaying with SSK modulation

2.4.1 System Model

We consider a bidirectional relay network, which is comprised of two source nodes, denoted by S_1 and S_2, and a relay node, denoted by R, as depicted in Fig. 2.14. We assume that both source nodes adopt SSK modulation such that the numbers of antennas at these nodes, N_{S_1} and N_{S_2}, are no less than 2, and the relay node with a single antenna serves as a non-regenerative repeater. The channel vectors for the $S_1 - R$ and $S_2 - R$ links, respectively, are represented by $\mathbf{h}$ and $\mathbf{g}$, where the magnitudes of the entries of $\mathbf{h}$ and $\mathbf{g}$, $|h_i|, i = 1, \ldots, N_{S_1}$ and $|g_l|, l = 1, \ldots, N_{S_2}$, are assumed to follow Nakagami-m distributions with parameters (Ω_h, m_h) and (Ω_g, m_g), respectively. We further presume that m_g and m_h are integers, the channels are mutually independent and reciprocal, and each node is subjected to AWGN of variance N_0. Each message exchange between two source nodes takes place in two phases. In the first phase, both sources simultaneously send the information and the following superimposed signal is received at the relay node

$$y_R = \sqrt{P_{S_1}} h_k + \sqrt{P_{S_2}} g_q + n_R, \tag{2.62}$$

where P_I is the transmit power of node I ($I \in \{S_1, S_2, R\}$), n_R is the AWGN at the relay, and without loss of generality, it is assumed that the k-th indexed antenna of S_1 and the q-th indexed antenna of S_2 are activated for transmission. In the second phase, the relay first processes the received signal by multiplying an amplification factor G and then forward it to the two source nodes. Due to symmetry, in what follows only the signal flowchart at S_1 will be detailed. The received signal at the

i-th antenna of S_1 can be expressed by

$$y_{S_1,i} = \sqrt{P_{S_1} P_R} G h_i h_k + \sqrt{P_{S_2} P_R} G h_i g_q + \sqrt{P_R} G h_i n_R + n_{S_1,i}, \tag{2.63}$$

where

$$G = \frac{1}{\sqrt{\left(P_{S_1} \Omega_h + P_{S_2} \Omega_g + N_0\right)}}, \tag{2.64}$$

is chosen to fix the average transmit power of relay and $n_{S_1,i}$ is the AWGN at the i-th antenna of S_1. Here, the fixed-gain AF relay mode is considered. We note that it is quite difficult to implement the CSI-assisted AF relay mode since the instantaneous CSI which carries on information due to SSK modulation is unavailable at the relay during the message transmission. Since the source acquires the knowledge of the activated antenna index for the previous transmission phase and the CSI during the pilot transmission phase, it is able to eliminate the self-interference as

$$\widetilde{y}_{S_1,i} = \sqrt{P_{S_2} P_R} G h_i g_q + \underbrace{\sqrt{P_R} G h_i n_R + n_{S_1,i}}_{Effective\ noise}. \tag{2.65}$$

Finally, based on the ML principle, the optimum detection for the transmit antenna index can be given by

$$\begin{aligned} \hat{q} &= \underset{1 \le q \le N_{S_2}}{\arg\max}\, f_{\widetilde{\mathbf{y}}_{S_1}}\left(\widetilde{\mathbf{y}}_{S_1} | g_q, \mathbf{h}\right) \\ &= \underset{1 \le q \le N_{S_2}}{\arg\min} \left(\widetilde{\mathbf{y}}_{S_1} - \mathbf{d}_{g_q}\right)^H \mathbf{C}^{-1}_{\mathbf{n}_{eff,S_1}} \left(\widetilde{\mathbf{y}}_{S_1} - \mathbf{d}_{g_q}\right), \end{aligned} \tag{2.66}$$

where

$$f_{\widetilde{\mathbf{y}}_{S_1}}\left(\widetilde{\mathbf{y}}_{S_1} | g_q, \mathbf{h}\right) = \pi^{-N_{S_1}} \left|\mathbf{C}_{\mathbf{n}_{eff,S_1}}\right|^{-\frac{1}{2}} e^{-\left(\widetilde{\mathbf{y}}_{S_1} - \mathbf{d}_{g_q}\right)^H \mathbf{C}^{-1}_{\mathbf{n}_{eff,S_1}} \left(\widetilde{\mathbf{y}}_{S_1} - \mathbf{d}_{g_q}\right)},$$

$\widetilde{\mathbf{y}}_{S_1}$, of dimension N_{S_1}, is the received signal vector, whose i-th entry is given by (2.65), $\mathbf{d}_{g_q}$, of dimension N_{S_1}, is the useful signal vector, whose i-th entry is $\sqrt{P_{S_2} P_R} G h_i g_q$, and $\mathbf{C}_{\mathbf{n}_{eff,S_1}}$, of dimensions $N_{S_1} \times N_{S_1}$, is the covariance matrix of the effective noise, whose (i, i')-th entry is $N_0 \left(P_R G^2 h_i^* h_{i'} + \delta\left(i - i'\right)\right)$.

As indicated in (2.66), the optimum ML detector invokes an inverse of a positive definite matrix $\mathbf{C}_{\mathbf{n}_{eff,S_1}}$, resulting in a cubic increase in the dimension of the matrix instead of the linear increase with the number of received antennas in previous SSK modulation schemes [17, 18]. This is due to the fact that by incurring the relay retransmission, the effective noise at different antennas of either source become highly correlated. To solve this problem, we propose a suboptimal detector, which enables not only a largely reduced detection but also a near-ML performance as will be validated in Sect. 2.4.3. The idea is motivated by the discovery of the underlying spatial constellation for SSK modulation in [17] and can be realized based on the MRC principle [9, pp. 821–823] as follows,

$$\hat{q} = \arg\min_{1 \le q \le N_{S_2}} \left| \frac{\mathbf{h}^H \widetilde{\mathbf{y}}_{S_1}}{\|\mathbf{h}\|_F^2} - g_q \right|. \tag{2.67}$$

It is clear that the evaluation of (2.67) requires $(4N_{S_1} + 2N_{S_2} - 1)$ complex operations, which is much less than the $\mathcal{O}\left(N_{S_2} N_{S_1}^3\right)$ required in (2.66). It is worth noting that the powers of the effective noise processes at all antennas are simply treated as equal while performing MRC in (2.67), which greatly helps relieve the burden on the source nodes in estimating the noise power at different antennas. Also note that unlike the MRC detection in [3], which only operates under special channel assumption, the one in (2.67) holds regardless of the channel realization.

2.4.2 ABEP Analysis and Power Allocation

In this subsection, we derive upper bounded and asymptotic ABEPs of the system. Based on the asymptotic ABEP, we further solve the power allocation problem.

PEP Derivation

The PEP Pr $\left(q \to \hat{q}\right)$, which is defined as the probability of the error event that occurs when the q-th antenna of S_2 is activated but antenna index $\hat{q}$ is detected, is derived in closed form in the following.

When conditioned upon the instantaneous CSI, the PEP can be readily expressed from (2.65) and (2.67) as

$$\begin{aligned}
\Pr\left(q \to \hat{q} | \mathbf{h}, \mathbf{g}\right) &= \Pr\left(\left|\mathbf{h}^H \widetilde{\mathbf{y}}_{S_1} / \|\mathbf{h}\|^2 - g_q\right|^2 > \left|\mathbf{h}^H \widetilde{\mathbf{y}}_{S_1} / \|\mathbf{h}\|^2 - g_{\hat{q}}\right|^2\right) \\
&= \Pr\left(\frac{\left|g_q - g_{\hat{q}}\right|^2}{2} < \Re\left\{\left(g_{\hat{q}} - g_q\right)^* n_{eq}\right\}\right) \\
&= Q\left(\sqrt{\frac{\gamma_{R-S_1} \cdot \gamma_{S_2-R}}{\gamma_{R-S_1} + C_{S_1}}}\right),
\end{aligned} \tag{2.68}$$

where $C_{S_1} = 1/(N_{S_1} G^2 N_0)$, $n_{eq} = n_R/\sqrt{P_{S_2}} + \mathbf{h}^H \mathbf{n}_{S_1}/(\sqrt{P_{S_2} P_R} G \|\mathbf{h}\|^2)$ is the equivalent noise after MRC, $\mathbf{n}_{S_1}$ is the noise vector at S_1, and $\gamma_{R-S_1} = P_R \|\mathbf{h}\|^2 / (N_{S_1} N_0)$ and $\gamma_{S_2-R} = P_{S_2} \left|g_q - g_{\hat{q}}\right|^2 / (2N_0)$ are defined as the instantaneous received SNR for the $R - S_1$ link and the $S_2 - R$ link, respectively. From (2.68), we see that to obtain the PEP dispensing with the channels, it is necessary to characterize the distributions of γ_{R-S_1} and γ_{S_2-R} first.

By considering an integer-fading-parameter Nakagami-m environment, the PDF of γ_{R-S_1} can be expressed from [19, Eq. (7)] as

$$f_{\gamma_{R-S_1}}(\gamma) = \frac{\gamma^{m_h N_{S_1}-1}}{\Gamma\left(m_h N_{S_1}\right)} \left(\frac{m_h N_{S_1}}{\overline{\gamma}_{R-S_1}}\right)^{m_h N_{S_1}} e^{-\frac{m_h N_{S_1}}{\overline{\gamma}_{R-S_1}}\gamma}, \tag{2.69}$$

where $\overline{\gamma}_{R-S_1} = \frac{P_R \Omega_h}{N_0}$ is the average received SNR for the $R - S_1$ link. On the other hand, since the PDF of $\left|g_q - g_{\hat{q}}\right|$ is given by [20, Eq. (8)] of a form

$$f_{\left|g_q - g_{\hat{q}}\right|}(u) = \wp_{\mathbf{g}}\left\langle \frac{m_g^{z+1}}{2^z \Omega_g^{z+1}} u^{2z+1} e^{-\frac{m_g u^2}{2\Omega_g}} \right\rangle, \tag{2.70}$$

where the operator $\wp_{\mathbf{g}}\langle\cdot\rangle$ is defined as

$$\wp_{\mathbf{g}}\langle\chi\rangle = \sum_{i_0=0}^{m_g-1}\sum_{i_1=0}^{m_g-1} \frac{(i_0+i_1)!(-m_g+1)_{i_0}(-m_g+1)_{i_1}}{(i_0!)^2(i_1!)^2 2^{i_0+i_1}} \sum_{z=0}^{i_0+i_1} \frac{(-i_0-i_1)_z}{(z!)^2}\chi, \tag{2.71}$$

from basic probability theory [9, p. 30–32] we have

$$\begin{aligned} f_{\gamma_{S_2-R}}(\gamma) &= \sqrt{\frac{N_0}{2P_{S_2}\gamma}} f_{\left|g_q - g_{\hat{q}}\right|}\left(\sqrt{\frac{2N_0}{P_{S_2}}\gamma}\right) \\ &= \wp_{\mathbf{g}}\left\langle \left(\frac{m_g}{\overline{\gamma}_{S_2-R}}\right)^{z+1} \gamma^z e^{-\frac{m_g}{\overline{\gamma}_{S_2-R}}\gamma} \right\rangle, \end{aligned} \tag{2.72}$$

where $\overline{\gamma}_{S_2-R} = \frac{P_{S_2}\Omega_g}{N_0}$ is the average received SNR for the $S_2 - R$ link. Accordingly, given the PDF in (2.72), the cumulative distribution function (CDF) of γ_{S_2-R} can be readily derived as

$$\begin{aligned} F_{\gamma_{S_2-R}}(\gamma) &= \wp_{\mathbf{g}}\left\langle \Gamma(z+1) - \Gamma\left(z+1, \frac{m_g}{\overline{\gamma}_{S_2-R}}\gamma\right) \right\rangle \\ &= 1 - \wp_{\mathbf{g}}\left\langle z! e^{-\frac{m_g}{\overline{\gamma}_{S_2-R}}\gamma} \sum_{m=0}^{z} \left(\frac{m_g}{\overline{\gamma}_{S_2-R}}\right)^m \frac{\gamma^m}{m!} \right\rangle, \end{aligned} \tag{2.73}$$

where the property $\wp_{\mathbf{g}}\langle\Gamma(z+1)\rangle = 1$ and the series representation of the incomplete gamma function in [12, Eq. (8.352.2)] are used in obtaining the last equation.

Therefore, by analogy with the mathematics in [18], the CDF of the equivalent received SNR in (2.68), i.e., $\gamma_{eq} = \frac{\gamma_{R-S_1}\cdot\gamma_{S_2-R}}{\gamma_{R-S_1}+C_{S_1}}$, can be readily derived as

$$\begin{aligned}F_{\gamma_{eq}}(\gamma) &= \int_0^{+\infty} F_{\gamma_{S_2-R}}\left(\left(1+\frac{C}{x}\right)\gamma\right) f_{\gamma_{R-S_1}}(x)\,dx\\ &=1-\frac{e^{-\frac{m_g}{\bar{\gamma}_{S_2-R}}\gamma}}{\Gamma\left(m_h N_{S_1}\right)}\left(\frac{m_h N_{S_1}}{\bar{\gamma}_{R-S_1}}\right)^{m_h N_{S_1}} \wp_{\mathbf{g}}\left\langle z!\sum_{m=0}^{z}\frac{\gamma^m}{m!}\left(\frac{m_g}{\bar{\gamma}_{S_2-R}}\right)^m\right.\\ &\quad\left.\times\int_0^{+\infty} x^{m_h N_{S_1}-1}\left(1+\frac{C}{x}\right)^m e^{-\frac{m_g C\gamma}{\bar{\gamma}_{S_2-R}}\cdot\frac{1}{x}-\frac{m_h N_{S_1}}{\bar{\gamma}_{R-S_1}}x}dx\right\rangle.\end{aligned}\tag{2.74}$$

Next, by resorting to the binomial expansion of $\left(1+\frac{C}{x}\right)^m$ and the integral result in [12, Eq. (3.471.9)], (2.74) can be further simplified as

$$\begin{aligned}F_{\gamma_{eq}}(\gamma) =&1-\frac{2e^{-\frac{m_g}{\bar{\gamma}_{S_2-R}}\gamma}}{\Gamma\left(m_h N_{S_1}\right)}\wp_{\mathbf{g}}\left\langle\sum_{m=0}^{z}\sum_{n=0}^{m}\frac{z!}{(m-n)!n!}D_{S_1}^{\frac{m_h N_{S_1}+n}{2}}\right.\\ &\left.\times\left(\frac{m_g}{\bar{\gamma}_{S_2-R}}\right)^{m-n}\gamma^{\frac{m_h N_{S_1}-n}{2}+m} K_{m_h N_{S_1}-n}\left(2\sqrt{D_{S_1}\gamma}\right)\right\rangle,\end{aligned}\tag{2.75}$$

where $D_{S_1}=\frac{m_g m_h C_{S_1} N_{S_1}}{\bar{\gamma}_{S_2-R}\bar{\gamma}_{R-S_1}}=m_g m_h\frac{\bar{\gamma}_{S_1-R}+\bar{\gamma}_{S_2-R}+1}{\bar{\gamma}_{S_2-R}\bar{\gamma}_{R-S_1}}$ and $\bar{\gamma}_{S_1-R}=\frac{P_{S_1}\Omega_h}{N_0}$ is the average received SNR for the S_1-R link. Finally, by using the partial integral, the PEP can be expressed as

$$\begin{aligned}\Pr\left(q\to\hat{q}\right) &= \frac{1}{2\sqrt{2\pi}}\int_0^{+\infty}\frac{1}{\sqrt{\gamma}}e^{-\frac{\gamma}{2}}F_{\gamma_{eq}}(\gamma)\,d\gamma\\ &=\frac{1}{2}-\frac{1}{\sqrt{2\pi}\Gamma\left(m_h N_{S_1}\right)}\wp_{\mathbf{g}}\left\langle\sum_{m=0}^{z}\sum_{n=0}^{m}\frac{z!}{(m-n)!n!}D_{S_1}^{\frac{m_h N_{S_1}+n}{2}}\left(\frac{m_g}{\bar{\gamma}_{S_2-R}}\right)^{m-n}\right.\\ &\quad\left.\times\int_0^{+\infty}\gamma^{\frac{m_h N_{S_1}-n-1+2m}{2}}e^{-\left(\frac{m_g}{\bar{\gamma}_{S_2-R}}+\frac{1}{2}\right)\gamma}K_{m_h N_{S_1}-n}\left(2\sqrt{D_{S_1}\gamma}\right)d\gamma\right\rangle.\end{aligned}\tag{2.76}$$

Closed-form solution exists by changing the integral variable with $t=\sqrt{\gamma}$ in the last equation of (2.76) and applying the identity in [12, Eq. (6.631.3)], yielding

$$\begin{aligned}\Pr\left(q\to\hat{q}\right) =&\frac{1}{2}-\frac{e^{D_{S_1}/\left(\frac{2m_g}{\bar{\gamma}_{S_2-R}}+1\right)}}{2\sqrt{2\pi D_{S_1}}\Gamma\left(m_h N_{S_1}\right)}\wp_{\mathbf{g}}\left\langle\sum_{m=0}^{z}\sum_{n=0}^{m}\frac{z!D_{S_1}^{\frac{m_h N_{S_1}+n}{2}}}{(m-n)!n!}\right.\\ &\times\left(\frac{m_g}{\bar{\gamma}_{S_2-R}}\right)^{m-n}\left(\frac{m_g}{\bar{\gamma}_{S_2-R}}+\frac{1}{2}\right)^{-\frac{m_h N_{S_1}-n+2m}{2}}\Gamma\left(\frac{1}{2}+m_h N_{S_1}+m-n\right)\\ &\left.\times\,\Gamma\left(\frac{1}{2}+m\right)W_{-\frac{m_h N_{S_1}-n+2m}{2},\frac{m_h N_{S_1}-n}{2}}\left(\frac{D_{S_1}}{\frac{m_g}{\bar{\gamma}_{S_2-R}}+\frac{1}{2}}\right)\right\rangle,\end{aligned}\tag{2.77}$$

which with the help of [21, Eqs. (13.1.10) and (13.1.33)] can be also expressed in terms of ${}_2F_0$ as

$$\Pr\left(q \to \hat{q}\right) = \frac{1}{2} - \frac{\Gamma^{-1}\left(m_h N_{S_1}\right)}{2\sqrt{2\pi}} \wp_{\mathbf{g}} \left\langle \sum_{m=0}^{z} \sum_{n=0}^{m} \frac{z! D_{S_1}^{n-m-\frac{1}{2}}}{(m-n)!n!} \right.$$
$$\times \left(\frac{m_g}{\bar{\gamma}_{S_2-R}}\right)^{m-n} \Gamma\left(\frac{1}{2} + m_h N_{S_1} + m - n\right) \Gamma\left(\frac{1}{2} + m\right)$$
$$\left. \times {}_2F_0\left(m_h N_{S_1} + m - n + \frac{1}{2}, m + \frac{1}{2};; -\left(\frac{m_g}{D_{S_1}\bar{\gamma}_{S_2-R}} + \frac{1}{2D_{S_1}}\right)\right)\right\rangle. \tag{2.78}$$

For a Rayleigh fading environment, where $m_g = m_h = 1$, (2.78) reduces to

$$\Pr\left(q \to \hat{q}\right) = \frac{1}{2} - \frac{\Gamma\left(N_{S_1} + \frac{1}{2}\right)}{4\Gamma\left(N_{S_1}\right)} \sqrt{\frac{2\bar{\gamma}_{R-S_1}\bar{\gamma}_{S_2-R}}{\bar{\gamma}_{S_1-R} + \bar{\gamma}_{S_2-R} + 1}}$$
$$\times {}_2F_0\left(N_{S_1} + \frac{1}{2}, \frac{1}{2};; -\frac{\bar{\gamma}_{R-S_1} + \frac{1}{2}\bar{\gamma}_{R-S_1}\bar{\gamma}_{S_2-R}}{\bar{\gamma}_{S_1-R} + \bar{\gamma}_{S_2-R} + 1}\right). \tag{2.79}$$

Upper-Bounded and Asymptotic ABEPs

When $N_{S_2} = 2$, one can readily determine that the exact ABEP at S_1, i.e., P_{b,S_1}, turns out to be (2.78). As for $N_{S_2} > 2$, the union bounding technique is used here to derive an upper bound on the ABEP at S_1:

$$P_{b,S_1} \leq \frac{1}{N_{S_2}\log_2\left(N_{S_2}\right)} \sum_{q=1}^{N_{S_2}} \sum_{\hat{q}=q+1}^{N_{S_2}} 2N\left(q, \hat{q}\right) \Pr\left(q \to \hat{q}\right) \tag{2.80}$$

where $N\left(q, \hat{q}\right)$ is the Hamming distance between the bit sequences mapped to antenna indexes q and $\hat{q}$. In the case in which N_{S_2} is equal to a power of 2, which is of great interest in practice, it follows that $\sum_{q=1}^{N_{S_2}} \sum_{\hat{q}=q+1}^{N_{S_2}} 2N\left(q, \hat{q}\right) = \frac{1}{2} N_{S_2}^2 \log_2\left(N_{S_2}\right)$ and (2.80) becomes [22]

$$P_{b,S_1} \leq \frac{N_{S_2}}{2} \Pr\left(q \to \hat{q}\right). \tag{2.81}$$

Furthermore, greater insight ABEP about the effects of the different system parameters can be gained from the following asymptotic expression for the ABEP:

$$P_{b,S_1} \approx \frac{m_g N_{S_2}}{4} \left(1 - \wp_{\mathbf{g}}\left\langle z! U\left(z-1\right)\right\rangle\right)$$
$$\times \left(\frac{1}{\bar{\gamma}_{S_2-R}} + \frac{m_h}{m_h N_{S_1} - 1} \frac{\bar{\gamma}_{S_1-R} + \bar{\gamma}_{S_2-R} + 1}{\bar{\gamma}_{R-S_1}\bar{\gamma}_{S_2-R}}\right). \tag{2.82}$$

Proof According to [23], the asymptotic error probability is closely related to the first non-zero higher order derivative of the CDF of γ_{eq} at the origin, which is contained in the first term of the Taylor's series of this CDF. By using the series representation of $K_{m_h N_{S_1}-n}\left(2\sqrt{D_{S_1}\gamma}\right)$ in (2.75) according to [12, Eq. (8.446)], one can determine that two parts in (2.75) contribute to the aforementioned first term, where the first part corresponds to the case when $m = 0$, given by

$$\begin{aligned} F^{(1)}_{\gamma_{eq}}(\gamma) =& 1 - \frac{2}{\Gamma\left(m_h N_{S_1}\right)}\left(1 - \frac{m_g}{\bar{\gamma}_{S_2-R}}\gamma + O(\gamma)\right) \\ &\times \left(\frac{\Gamma\left(m_h N_{S_1}\right)}{2} - \frac{\Gamma\left(m_h N_{S_1}-1\right)}{2} D_{S_1}\gamma + O(\gamma)\right) \\ =& \left(\frac{m_g}{\bar{\gamma}_{S_2-R}} + \frac{D_{S_1}}{m_h N_{S_1}-1}\right)\gamma + O(\gamma), \end{aligned} \quad (2.83)$$

and the second part corresponds to the case when $m = 1$, yielding

$$\begin{aligned} F^{(2)}_{\gamma_{eq}}(\gamma) =& -\frac{2\left(1 - \frac{m_g}{\bar{\gamma}_{S_2-R}}\gamma + O(\gamma)\right)}{\Gamma\left(m_h N_{S_1}\right)} \left[\wp_{\mathbf{g}}\langle z!U(z-1)\rangle \right. \\ &\times \left.\left(\frac{m_g \Gamma\left(m_h N_{S_1}\right)}{2\bar{\gamma}_{S_2-R}} + \frac{\Gamma\left(m_h N_{S_1}-1\right)}{2} D_{S_1}\right)\gamma + O(\gamma)\right] \\ =& -\wp_{\mathbf{g}}\langle z!U(z-1)\rangle \left(\frac{m_g}{\bar{\gamma}_{S_2-R}} + \frac{D_{S_1}}{m_h N_{S_1}-1}\right)\gamma + O(\gamma). \end{aligned} \quad (2.84)$$

Then, by incorporating (2.83) and (2.84), we obtain the first non-zero higher order derivative of $F_{\gamma_{eq}}(\gamma)$ at the origin as

$$\begin{aligned} \frac{\partial F_{\gamma_{eq}}(\gamma=0)}{\partial\gamma} &= f_{\gamma_{eq}}(0) \\ &= \left(1 - \wp\langle z!U(z-1)\rangle\right)\left(\frac{m_g}{\bar{\gamma}_{S_2-R}} + \frac{D_{S_1}}{m_h N_{S_1}-1}\right), \end{aligned} \quad (2.85)$$

where $f_{\gamma_{eq}}(\gamma)$ is the PDF of γ_{eq}. Finally, the application of the identity in [23, Eq. (10)] completes the proof. □

Specifically, for a Rayleigh fading environment, we have

$$P_{b,S_1} \approx \frac{N_{S_2}}{4}\left(\frac{1}{\bar{\gamma}_{S_2-R}} + \frac{1}{N_{S_1}-1}\,\frac{\bar{\gamma}_{S_1-R} + \bar{\gamma}_{S_2-R} + 1}{\bar{\gamma}_{R-S_1}\bar{\gamma}_{S_2-R}}\right). \quad (2.86)$$

From (2.82) and (2.86), it is clear that an increase in the number of received antennas in two-way relay systems with SSK modulation always achieves unit diversity order but results in an improvement of the coding gain. This is different from the case of conventional SSK modulation, where N_{S_1} diversity order can be gained [17]. When relay retransmission is introduced, the received signal versions at all antennas become dependent and completely unresolvable.

Power Allocation

By excluding all parameters irrelevant to power allocation (PA) and after some manipulation, the PA problem based on the minimization of the average ABEP $P_e = \left(P_{e,S_1} + P_{e,S_2}\right)/2$, subject to total and individual power constraints, can be formulated as

$$\begin{aligned}\left\{P_{S_1}^*, P_R^*, P_{S_2}^*\right\} = \min_{\left\{P_{S_1}, P_R, P_{S_2}\right\}} &\left\{\frac{1 - \wp_{\mathbf{g}}\langle z!U(z-1)\rangle}{m_h N_{S_1} \Omega_g}\right. \\ &\times \left(\frac{1}{P_{S_2}} + \frac{m_h}{m_h N_{S_1} - 1}\frac{P_{S_1}\Omega_h + P_{S_2}\Omega_g}{P_R P_{S_2}\Omega_h}\right) \\ &+ \frac{1 - \wp_{\mathbf{h}}\langle z!U(z-1)\rangle}{m_g N_{S_2}\Omega_h} \\ &\left.\times \left(\frac{1}{P_{S_1}} + \frac{m_g}{m_g N_{S_2} - 1}\frac{P_{S_1}\Omega_h + P_{S_2}\Omega_g}{P_R P_{S_1}\Omega_g}\right)\right\} \\ s.t. \quad P_{S_1} + P_R + P_{S_2} \leq P_{tot} \quad \& \quad P_{S_1}, P_R, P_{S_2} > 0, \end{aligned} \tag{2.87}$$

where P_{e,S_2} is the ABEP at S_2, $\wp_{\mathbf{h}}\langle\cdot\rangle$ is the operator defined similar to (2.71) but with all related parameters extracted from the $S_1 - R$ link, P_{tot} is the upper bound on the system power, and $\left\{P_{S_1}^*, P_R^*, P_{S_2}^*\right\}$ are the optimum values of $\left\{P_{S_1}, P_R, P_{S_2}\right\}$, which satisfy the constrained optimization problem in (2.87). A careful inspection of (2.87) reveals the convex property of the objective function with respect to the decision variables, i.e., P_{S_1}, P_R, and P_{S_2}, and thus unique solutions, i.e., $\left\{P_{S_1}^*, P_R^*, P_{S_2}^*\right\}$, exist for the above non-linear programming problem. Unfortunately, to the best of our knowledge, closed-form expressions for $\left\{P_{S_1}^*, P_R^*, P_{S_2}^*\right\}$ are unavailable. To this end, numerical searching algorithms should be relied on, and this can be easily realized by standard mathematical software platforms, e.g., Matlab, Mathematica, etc.

2.4.3 Performance Evaluation

In this subsection, BER simulations are conducted to validate the analysis given in Sect. 2.4.2. Without loss of generality, in the simulations we let $\Omega_h = \Omega_g = 1$ and the number of antennas at both sources be powers of 2. Unless otherwise specified, the transmit powers of all three nodes are set equal and the proposed MRC detector is applied.

Figure 2.15 depicts the BER performance for two different channel scenarios, where we choose $N_{S_1} = 8$ and $N_{S_2} = 2$. It is clear that the proposed MRC detector expresses nearly the same performance as the optimum ML detector. Moreover,

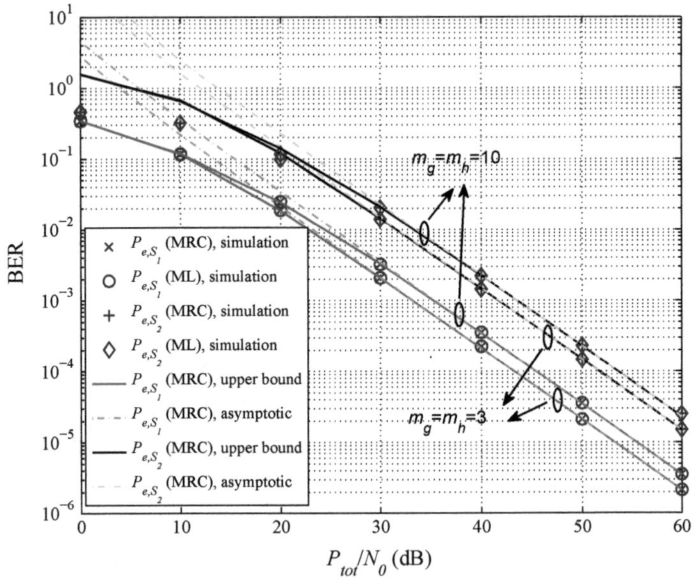

Fig. 2.15 BER performance in the case of $N_{S_1} = 8$ and $N_{S_2} = 2$

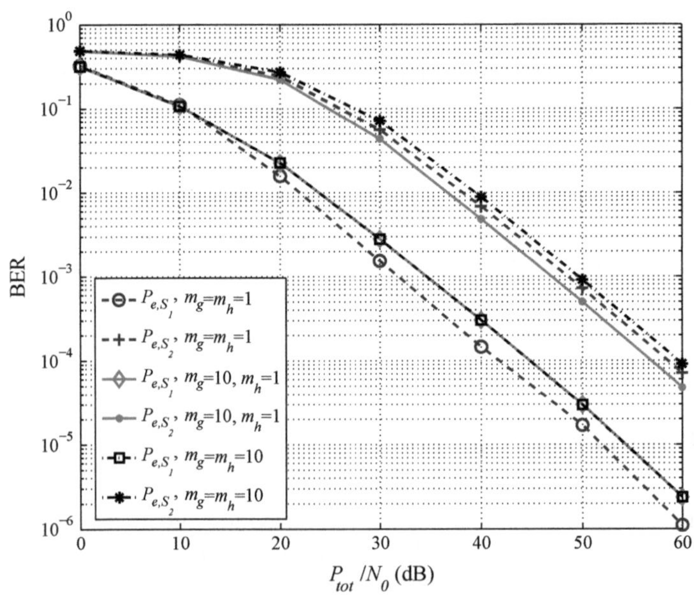

Fig. 2.16 Effects of fading parameters on the BER performance

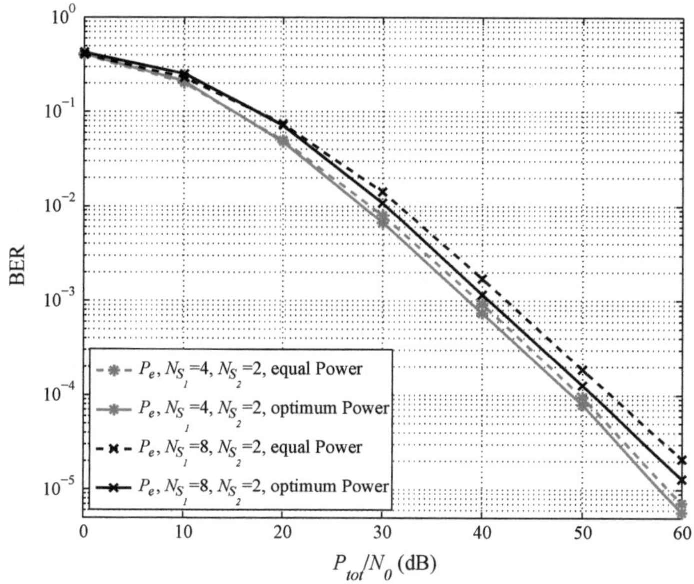

Fig. 2.17 BER performance comparison between equal power allocation and optimum power allocation

as expected, the analytical exact, upper bounded, and asymptotic ABEPs are all in perfect agreement with their simulation counterparts. On the other hand, as predicted in (2.82), the systems always gain a diversity order of unity.

Figure 2.16 shows the effects of fading parameters on the BER performance, where we choose $N_{S_1} = 32$ and $N_{S_2} = 2$. For the purpose of figure clarity, the analytical curves are removed. We see that as the fading of the link between one source and the relay becomes less severe, i.e., the value of the corresponding parameter $m_{h/g}$ becomes larger, the BER performance of this source will be somewhat improved, whereas that of the other source will deteriorate. The improvement in the BER performance is due to the improvement of the channel quality, which fulfils our intuition. However, deterioration of the BER performance seems counter-intuitive at first but can be explained reasonably after further consideration as follows: the improvement of the channel quality at one link will result in the shrinking of the spatial constellation for the source at this link, and thus will lead to the detection at the other source becoming harder. Moreover, from Fig. 2.16, it is clear that compared with the increase of the BER, the decrease of the BER is more serious. This can be easily accounted for from the positions of the two fading parameters in the BER expression in (2.82), which implies the dominant effect of the shrinking of the spatial constellation over the improvement of channel quality. Therefore, it is not surprising to see from Figs. 2.15 and 2.16 that the improvement of the fading severities for both links produces a reverse effect on the overall system performance, i.e., results in the deterioration of the overall system performance.

Figure 2.17 illustrates the feasibility of adopting the proposed PA strategy in lowering the average BER, where the fading parameters are fixed as $m_g = 1$ and $m_h = 10$. In Fig. 2.17, we consider two different system situations, i.e., (1) $N_{S_1} = 4$, $N_{S_2} = 2$; and (2) $N_{S_1} = 8$ and $N_{S_2} = 2$. The optimum PA strategies, according to (2.87), for the above two system settings in order are (1) $P_{S_1}^* = 0.4113 P_{tot}$, $P_R^* = 0.4317 P_{tot}$, $P_{S_2}^* = 0.1570 P_{tot}$; and (2) $P_{S_1}^* = 0.4398 P_{tot}$, $P_R^* = 0.4457 P_{tot}$, $P_{S_2}^* = 0.1144 P_{tot}$. Both BER values for the equal PA and optimum PA strategies are depicted in Fig. 2.17, while their analytical results are not shown for the purpose of figure clarity. One can observe that by adopting the proposed PA strategy, the system can even benefit about 2 dB SNR reduction with respect to the equal power distribution strategy at some specific level of P_e, e.g., $P_e = 10^{-3}$. We note that as the proposed PA strategy focuses on the minimization of average BER for a moderate-to-high SNR regime, it may worsen the average BER compared with the equal PA strategy for a very low SNR regime, where $P_e > 10^{-1}$.

2.5 Summary

In this chapter, we began with the introduction of the first space domain index modulation technique, which is well known as SM. Then, we proposed a receiver-side SM scheme, i.e., GPQSM, which relies on the ZF/MMSE pre-coding operation at the transmitter. The advantages of GPQSM were revealed through both theoretical analysis and simulations. However, similar to the other pre-coding SM schemes, the performance of GPQSM is limited by the correlated channels. To solve this problem, we further proposed the VSM scheme, which embeds additional information into the indices of the virtual parallel channels resulting from the SVD of the MIMO channel and can be regarded as either a transmitter-side or receiver-side SM scheme. VSM was shown to be favorable especially for low data rate transmission. Finally, we took SSK as a representative to examine the potential of space domain index modulation techniques in the application to cooperative communications.

References

1. N. Serafimovski, M. Renzo, S. Sinanovic, R. Mesleh, H. Haas, Fractional bit encoded spatial modulation (FBE-SM). IEEE Commun. Lett. **14**(5), 429–431 (2010)
2. J. Jeganathan, A. Ghrayeb, L. Szczecinski, Spatial modulation: optimal detection and performance analysis. IEEE Commun. Lett. **12**(8), 545–547 (2008)
3. R. Mesleh, H. Haas, S. Sinanović, C.W. Ahn, S. Yun, Spatial modulation. IEEE Trans. Veh. Tech. **57**(4), 2228–2241 (2008)
4. J. Wang, S. Jia, J. Song, Signal vector based detection scheme for spatial modulation. IEEE Commun. Lett. **16**(1), 19–21 (2012)
5. J. Zheng, Signal vector based list detection for spatial modulation. IEEE Wirel. Commun. Lett. **1**(4), 265–267 (2012)

6. Q. Tang, Y. Xiao, P. Yang, Q. Yu, S. Li, A new low-complexity near-ML detection algorithm for spatial modulation. IEEE Wirel. Commun. Lett. **2**(1), 90–93 (2013)
7. R. Rajashekar, K.V.S. Hari, L. Hanzo, Reduced-complexity ML detection and capacity-optimized training for spatial modulation systems. IEEE Trans. Commun. **62**(1), 112–125 (2014)
8. H. Men, M. Jin, A low-complexity ML Detection algorithm for spatial modulation systems with MPSK constellation. IEEE Commun. Lett. **18**(8), 1375–1375 (2014)
9. J.G. Proakis, *Digital Communications*, 3rd edn. (McGraw-Hill, New York, 1995)
10. D. Gore, R. Heath, A. Paulraj, Transmit selection in spatial multiplexing systems. IEEE Commun. Lett. **6**(11), 491–493 (2002)
11. M. Alouini, A. Abdi, M. Kaveh, Sum of gamma variates and performance of wireless communication systems over Nakagami-fading channels. IEEE Trans. Veh. Technol. **50**(6), 1471–1480 (2001)
12. I.S. Gradshteyn, I.M. Rhyhik, *Table of Integrals, Series and Products*, 7th edn. (Academic, New York, 2007)
13. W. Gifford, M. Win, M. Chiani, Diversity with practical channel estimation. IEEE Trans. Wirel. Commun. **4**(4), 1935–1947 (2005)
14. A.T. James, Distributions of matrix variates and latent roots derived from normal samples. Ann. Math. Stat. **35**, 475–501 (1964)
15. M. Chiani, D. Dardari, Improved exponential bounds and approximation for the Q-function with application to average error probability computation, in *Proceedings of IEEE Global Telecommunication Conference, Bologna, Italy* (2002), pp. 1399–1402
16. Z. Wang, G. Giannakis, A simple and general parameterization quantifying performance in fading channels. IEEE Trans. Commun. **51**(8), 1389–1398 (2003)
17. J. Jeganathan, A. Ghrayeb, L. Szczecinski, A. Ceron, Space shift keying modulation for MIMO channels. IEEE Trans. Wirel. Commun. **8**(7), 3692–3703 (2009)
18. R. Mesleh, S. Ikki, M. Alwakeel, Performance analysis of space shift keying with amplify and forward relaying. IEEE Commun. Lett. **15**(99), 1350–1352 (2011)
19. E.K. Al-Hussaini, A.A. Al-Bassiouni, Performance of MRC diversity systems for the detection of signals with Nakagami fading. IEEE Trans. Commun. **COM–33**(12), 1315–1319 (1985)
20. G.K. Karagiannidis, A closed-form solution for the distribution of the sum of Nakagami-*m* random phase vectors. IEEE Commun. Lett. **10**(12), 828–830 (2006)
21. M. Abramowitz, I.A. Stegun, *Handbook of Mathematical Functions With Formulas, Graphs, and Mathematical Tables*, 9th edn. (Dover, New York, 1964)
22. M. Di Renzo, H. Haas, Bit error probability of space-shift keying MIMO over multiple-access independent fading channels. IEEE Trans. Veh. Tech. **60**(8), 3694–3711 (2011)
23. A. Ribeiro, X. Cai, G.B. Giannakis, Symbol error probabilities for general cooperative links. IEEE Trans. Wirel. Commun. **4**(3), 1264–1273 (2005)

Chapter 3
Space-Time Domain Index Modulation

In this chapter, we first introduce a novel space-time domain index modulation scheme, named differential SM (DSM), which dispenses with the channel state information (CSI) at both the transmitter and receiver. Then, the design guidelines for performance improvement of DSM are provided. The pre-coding aided DSM (PDSM) is also presented, which is an extension of DSM to the receiver side. Finally, we investigate the application of DSM/PDSM to cooperative communications.

3.1 Differential Spatial Modulation

In this section, we develop DSM, which applies to any constant energy constellation such as PSK and to systems with arbitrary numbers of transmit and receive antennas while preserving the single active transmit antenna property. The ABEP performance of DSM is studied. With the same spectral efficiency, DSM is capable of paying no more than 3 dB SNR penalty compared with coherent SM and outperforming the single-antenna differential PSK (DPSK) and differential space-time coding schemes.

3.1.1 Transceiver Structure

Let the (m, t)-th entry of the $N_t \times N_t$ space-time block $\mathbf{S}$ denote the symbol s_{mt} transmitted via transmit antenna m at time instant t. In DSM, we require the $N_t \times N_t$ block $\mathbf{S}$ to satisfy the following conditions:

(1) Only one antenna remains active at each time instant; that is, only one entry in any column of $\mathbf{S}$ is nonzero.

M. Wen et al., *Index Modulation for 5G Wireless Communications*, Wireless Networks, DOI 10.1007/978-3-319-51355-3_3

(2) Each antenna is activated once and only once in the N_t successive time instants of a space-time block; that is, only one entry in any row of $\mathbf{S}$ is nonzero.
(3) The signal constellation is restricted to an equal energy M-PSK alphabet $\mathcal{X}$; that is, each nonzero entry is chosen from $\mathcal{X}$. A special case of $\mathcal{X}$ with $M = 1$ can be referred to as differential SSK (DSSK), in which all transmitted symbols are 1s.

An example of the space-time block for $N_t = 3$ is given by

$$\mathbf{S} = \begin{bmatrix} s_{11} & 0 & 0 \\ 0 & 0 & s_{23} \\ 0 & s_{32} & 0 \end{bmatrix}, \tag{3.1}$$

which means that at time instants 1, 2, and 3, the symbols s_{11}, s_{32}, and s_{23} drawn from $\mathcal{X}$ are sent from transmit antennas 1, 3, and 2, respectively, whereas the other two transmit antennas remain idle. Note that here we exploit the time domain by transmitting blocks of signals using an $N_t \times N_t$ transmission matrix. By this design, each antenna is activated once and only once during each block, thus making differential operation possible so long as the wireless channel remains unchanged over two consecutive blocks.

In order to facilitate differential (de-)modulation, one also has to make sure that the transmitted signal blocks satisfy the so-termed closure property (see e.g., [1]). Accordingly the received signal for DSM can be expressed in matrix form as

$$\mathbf{Y} = \mathbf{HS} + \mathbf{N}, \tag{3.2}$$

where $\mathbf{Y}$ is the received signal over N_t successive time instants with the entry y_{nt} denoting the received signal at receive antenna n at time instant t. $\mathbf{H}$ is the $N_r \times N_t$ channel matrix. The (n, m)-th entry h_{nm} denotes the path gain from transmit antenna m to receive antenna n which is complex-Gaussian with zero mean and unit variance, and independent across n and m. $\mathbf{N}$ is the $N_r \times N_t$ AWGN matrix. The entry n_{nt} denotes the noise received by receive antenna n at time instant t which is independent and identically distributed (over t and n) complex-Gaussian with zero mean and variance N_0. Since only one antenna remains active at any time instant, only one entry in the column t of $\mathbf{S}$ is nonzero, satisfying $E\{|s_{mt}|^2\} = 1$.

Let the set $\mathcal{G}$ contain all possible space-time blocks $\mathbf{S}$. Given the aforementioned three conditions, for any N_t, it can be readily proved that the set $\mathcal{G}$ holds the following closure property under multiplication:

$$\forall \mathbf{S}_1, \mathbf{S}_2 \in \mathcal{G}, \mathbf{S}_1\mathbf{S}_2 \in \mathcal{G}. \tag{3.3}$$

Here, we note that in order to enable the differential transmission, the number of time instants collected in a space-time block $\mathbf{S}$ has to be equal to N_t.

Spectral Efficiency of DSM

The set $\mathcal{G}$ with N_t transmit antennas and M-PSK constellation contains $N_t!M^{N_t}$ blocks. In theory, the spectral efficiency of the proposed DSM is (bps/Hz)

$$R_{DSM,\ theory} = \frac{1}{N_t}\log_2(N_t!) + \log_2(M). \tag{3.4}$$

Based on Stirling's Formula $n! \approx \sqrt{2\pi n}n^n e^{-n}$ [2], the spectral efficiency of DSM can be approximated as

$$R_{DSM,\ theory} \approx \log_2(N_t) - \underbrace{\left(\log_2(e) - \frac{1}{N_t}\log_2\left(\sqrt{2\pi N_t}\right)\right)}_{\textit{loss of spectral efficiency}} + \log_2(M), \tag{3.5}$$

where e is the mathematical constant. Note that the spectral efficiency of a conventional SM system is

$$R_{SM,\ theory} = \log_2(N_t) + \log_2(M). \tag{3.6}$$

Therefore, the loss of spectral efficiency of DSM compared to that of SM can be approximated as

$$\log_2(e) - \frac{1}{N_t}\log_2\left(\sqrt{2\pi N_t}\right) \le \log_2(e). \tag{3.7}$$

This indicates that the loss of the spectral efficiency is upper bounded by $\log_2(e)$.

Note that when $N_t \ge 3$, $N_t!$ is not an integer power of two. One way to approach the theoretical spectral efficiency $R_{DSM,\ theory}$ is to apply the so-called FBE to DSM, which was applied to SM for the case in which N_t is not an integer power of two. However, FBE leads to longer decoding delay and worse BER performance.

In practical use of SM, we often assume that N_t is an integer power of 2. Hence the spectral efficiency of SM is

$$\overline{R}_{SM} = \lfloor\log_2(N_t)\rfloor + \log_2(M). \tag{3.8}$$

Similarly, a simpler DSM without FBE is preferred. For this purpose, we choose $2^{\lfloor\log_2(N_t!)\rfloor}M^{N_t}$ transmit blocks from $\mathcal{G}$ and form a subset $\mathcal{G}_T$ for (de-)mapping. Each transmit block can then be encoded with $\lfloor\log_2(N_t!)\rfloor + N_t\log_2(M)$ bits. Therefore, the spectral efficiency of DSM for practical use can be written as

$$\overline{R}_{DSM} = \frac{1}{N_t}\lfloor\log_2(N_t!)\rfloor + \log_2(M). \tag{3.9}$$

In Fig. 3.1, we set $M = 1$, and compare the spectral efficiencies of SM and DSM in theory and for practical use. According to (3.4), (3.6), (3.8), and (3.9), we calculated the spectral efficiency of points with integral numbers of transmit antennas and connected the markers of each equation with a line. Note that the practical spectral efficiencies of SM and DSM behave very differently. The practical SM spectral efficiency is a piece-wise constant curve, whereas that of the practical DSM is rather

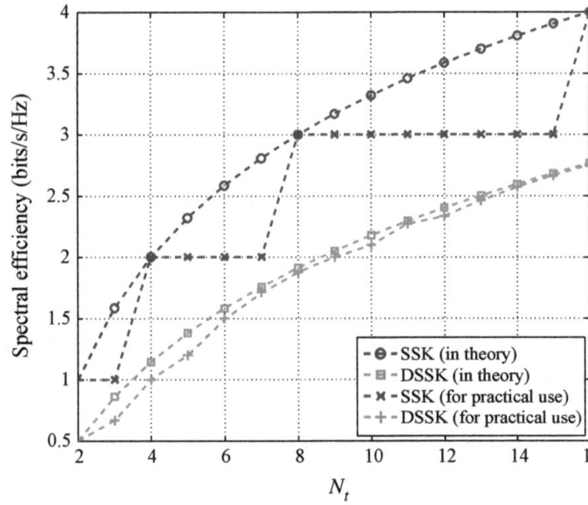

Fig. 3.1 Spectral efficiency comparison of SM and DSM when $M = 1$, i.e., SSK and DSSK, respectively, where SSK and DSSK for practical use refer to the schemes without FBE

"continuous." In other words, when N_t increases from 4 to 5, 6 and 7, the spectral efficiency of practical SM remains a constant, while that of practical DSM keeps increasing. This is because, by extending into the time domain, the allowable spatial modulation size increases as $N_t!$ for DSM, as opposed to N_t for SM. This is actually working to our advantage since, as shown in Fig. 3.1, the spectral efficiency gap between coherent and differential SM is much narrower in practical use than in theory. In fact, as can be seen from the figure, the practical DSM and the theoretical DSM share very similar spectral efficiencies. Considering its simplicity and lack of extra processing delay or error propagation, we adopt this encoding method.

Differential Transmission Process

The transmitter begins the transmission by sending an arbitrary initial block $\mathbf{S}_0 \in \mathcal{G}$. Since the M-PSK constellation $\mathcal{X}$ always contains symbol 1, we choose $\mathbf{S}_0 = \mathbf{I}_{N_t}$ for any N_t without loss of generality. Thereafter, the transmitter encodes the bit stream in a recursive manner. Suppose that from time instant $N_t(t-1)$ to $N_t t - 1$, the $(t-1)$-th block $\mathbf{S}_{t-1}$ is transmitted ($t = 1, 2, 3, \cdots$). This encoding process in the next space-time block operates as follows:

(1) *Step 1*: Given the input bit stream, we map $\lfloor \log_2(N_t!) \rfloor + N_t b$ bits onto $\mathcal{G}_T$ and obtain $\mathbf{X}_t \in \mathcal{G}_T \subseteq \mathcal{G}$.
Among all the possible mapping methods, without loss of generality, we choose the following mapping. The first $\lfloor \log_2(N_t!) \rfloor$ bits are mapped onto the spatial domain, determining the order of transmit antennas being activated in a space-time block by an index-mapping procedure (two different index-mapping procedures will be presented later). Then, the remaining $N_t b$ bits are mapped to the signal domain. Specifically, each group of b bits is mapped to a symbol drawn from the M-PSK constellation $\mathcal{X}$ for all N_t time instants. Taking into account the mapping results of both parts, we can determine $\mathbf{X}_t$.

(2) *Step 2*: Compute the transmitted block $\mathbf{S}_t$ as follows:

$$\mathbf{S}_t = \mathbf{S}_{t-1}\mathbf{X}_t. \tag{3.10}$$

Note that given $\mathbf{S}_{t-1}$ and $\mathbf{X}_t \in \mathcal{G}$, according to the property in Sect. II, it is guaranteed that $\mathbf{S}_t \in \mathcal{G}$.

(3) *Step 3*: Send the block $\mathbf{S}_t$ during time period $N_t t$ to $N_t(t+1)-1$.

The transmitter repeats the above three-step process till the end of the transmission. A block diagram of the transmitter is depicted in Fig. 3.2. An example of the differential transmission process with $M = 2$ and $N_t = 2$ is presented in Fig. 3.2 as well. In this case, each block can convey three bits. The first bit is encoded in the spatial domain, e.g., bit 0 determines a diagonal matrix and bit 1 determines an anti-diagonal matrix. The remaining two bits are encoded in the signal domain as two BPSK symbols.

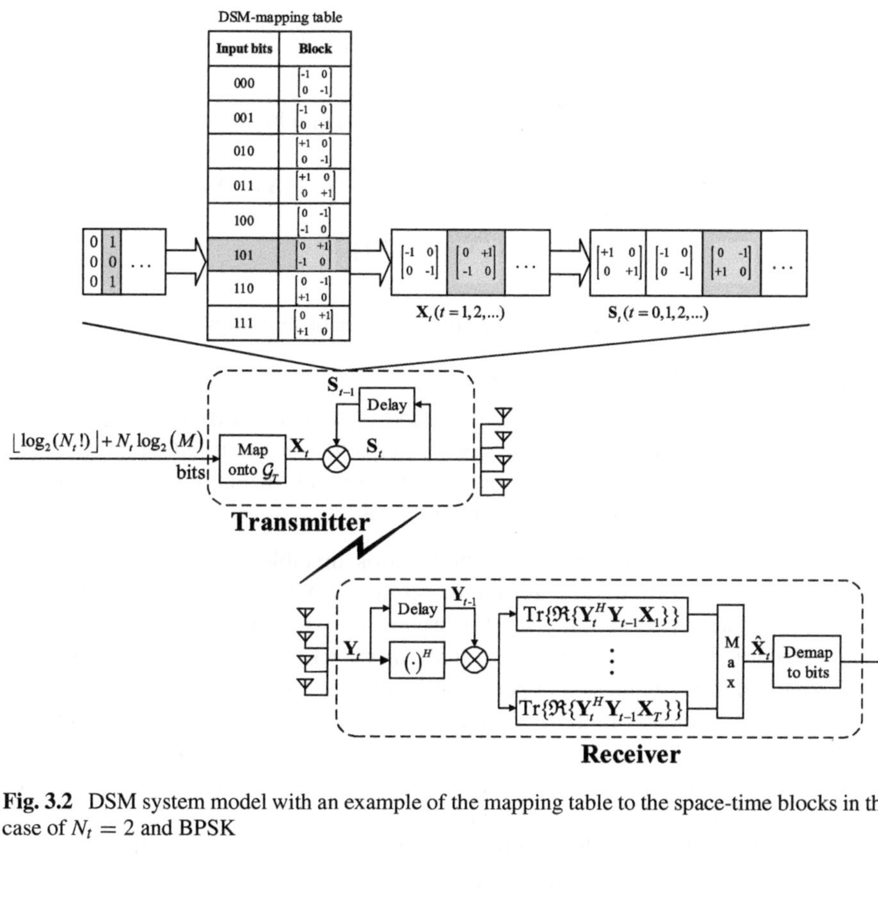

Fig. 3.2 DSM system model with an example of the mapping table to the space-time blocks in the case of $N_t = 2$ and BPSK

Differential Detection

Suppose that the $(t-1)$-th and the t-th received signal blocks are $\mathbf{Y}_{t-1}$ and $\mathbf{Y}_t$, respectively. From (3.2), we have

$$\mathbf{Y}_{t-1} = \mathbf{H}_{t-1}\mathbf{S}_{t-1} + \mathbf{N}_{t-1}, \tag{3.11}$$

and

$$\mathbf{Y}_t = \mathbf{H}_t\mathbf{S}_t + \mathbf{N}_t. \tag{3.12}$$

Assume a quasi-static channel in which the fading coefficients remain constant over two adjacent DSM transmit blocks, namely $2N_t$ symbol durations. By substituting (3.10) and (3.11) into (3.12), we have

$$\mathbf{Y}_t = \mathbf{Y}_{t-1}\mathbf{X}_t - \mathbf{N}_{t-1}\mathbf{X}_t + \mathbf{N}_t. \tag{3.13}$$

Therefore, the optimal ML detector can be derived as

$$\widehat{\mathbf{X}}_t = \arg\min_{\mathbf{X}} \parallel \mathbf{Y}_t - \mathbf{Y}_{t-1}\mathbf{X} \parallel^2 . \tag{3.14}$$

Applying the identity that $\mathrm{Tr}\{\mathbf{AB}\} = \mathrm{Tr}\{\mathbf{BA}\}$, (3.14) can be reduced to

$$\begin{aligned}\widehat{\mathbf{X}}_t &= \arg\min_{\mathbf{X}} \mathrm{Tr}\left\{(\mathbf{Y}_t - \mathbf{Y}_{t-1}\mathbf{X})^H (\mathbf{Y}_t - \mathbf{Y}_{t-1}\mathbf{X})\right\} \\ &= \arg\max_{\mathbf{X}} \mathrm{Tr}\left\{\Re\left\{\mathbf{Y}_t^H\mathbf{Y}_{t-1}\mathbf{X}\right\}\right\}.\end{aligned} \tag{3.15}$$

According to (3.15), a block diagram of the receiver is shown in Fig. 3.2.

Implementation of the Index-Mapping

We focus on the index-mapping procedure and provide two different implementation methods. The index-mapping procedure maps the incoming bits to the order of transmit antennas being activated in a space-time block, which can be denoted by the permutation of N_t indices of transmit antennas.

(1) Look-Up Table Method: In this method, a look-up table is created to provide the corresponding permutations for the incoming $\lfloor \log_2(N_t!) \rfloor$ bits. An example for $N_t = 3$ is presented in Table 3.1. Since $3! = 6$, two permutations out of the six are discarded. The selection of permutations may affect the overall performance. For ease of implementation of permutation index-mapping, we choose to discard the lexicographically larger permutations in this development. For arbitrary N_t, the size of the look-up table is $2^{\lfloor \log_2(N_t!) \rfloor}$. This is an efficient and simple method when the size of the table is small. However, it is not feasible for larger values of N_t as the table size grows exponentially.
(2) Permutation Method: This method introduces a one-to-one mapping between integers and permutations of N_t elements in lexicographical order, i.e., it maps

Table 3.1 An example of look-up table for $N_t = 3$

Bits	Antenna activation order	Blocks
00	(1 2 3)	$\begin{bmatrix} s_{11} & 0 & 0 \\ 0 & s_{22} & 0 \\ 0 & 0 & s_{33} \end{bmatrix}$
01	(1 3 2)	$\begin{bmatrix} s_{11} & 0 & 0 \\ 0 & 0 & s_{23} \\ 0 & s_{32} & 0 \end{bmatrix}$
10	(2 1 3)	$\begin{bmatrix} 0 & s_{12} & 0 \\ s_{21} & 0 & 0 \\ 0 & 0 & s_{33} \end{bmatrix}$
11	(2 3 1)	$\begin{bmatrix} 0 & 0 & s_{13} \\ s_{21} & 0 & 0 \\ 0 & s_{32} & 0 \end{bmatrix}$

an integer m to a sequence $\mathbf{a}^{(m)} = (a_1^{(m)}, \cdots, a_{N_t}^{(m)})$ which is a permutation of the set $\{1, \cdots, N_t\}$. For fixed N_t, all $m \in [0, N_t! - 1]$ can be presented by a permutation $\mathbf{a}^{(m)}$ of length N_t. This mapping is the so-called Lehmer code [3]. Application of this method to DSM is presented as follows:

(i) *Convert the integer m to its factorial representation $\mathbf{b}^{(m)}$ of length N_t.* We define the factorial sequence $\mathbf{b}^{(m)} = (b_1^{(m)}, \cdots, b_{N_t}^{(m)})$, which takes elements according to the following equation:

$$m = b_1^{(m)}(N_t - 1)! + \cdots + b_{N_t}^{(m)} 0!. \tag{3.16}$$

As an example, for $N_t = 4$ and $m = 11$, the following factorial sequence $\mathbf{b}^{(m)}$ can be calculated:

$$\begin{aligned} 11 &= 1 \cdot 3! + 2 \cdot 2! + 1 \cdot 1! + 0 \cdot 0! \\ &\rightarrow \mathbf{b}^{(11)} = (1, 2, 1, 0). \end{aligned} \tag{3.17}$$

The algorithm, which calculates the factorial sequence $\mathbf{b}^{(m)}$, starts by choosing the maximal $b_1^{(m)}$ that satisfies $b_1^{(m)}(N_t - 1)! \leq m$ and then chooses the maximal $b_2^{(m)}$ that satisfies $b_2^{(m)}(N_t - 2)! \leq m - b_1^{(m)}(N_t - 1)!$, and so on.

(ii) *Map the factorial sequence $\mathbf{b}^{(m)}$ to a permutation $\mathbf{a}^{(m)}$.* We define $\Theta = (1, 2, \cdots, N_t)$ as an ordered list with its first element zero indexed. The leftmost permutation element $a_1^{(m)}$ is chosen as $\Theta_{b_1^{(m)}}$ and then the element $\Theta_{b_1^{(m)}}$ is removed from the list $\Theta = (1, 2, \cdots, N_t)$. Think of this new list as zero indexed and one can obtain each successive element of $\mathbf{a}^{(m)}$ in a recursive manner:

For $l = 1 : N_t$

$a_l^{(m)} = \Theta_{b_l^{(m)}}$

$\Theta_{b_l^{(m)}}$ is removed from Θ

End For

As an example with $N_t = 4$ and $m = 11$, from the above procedure we have $\mathbf{b}^{(11)} = (1, 2, 1, 0)$. Letting $\Theta = (1, 2, 3, 4)$, we have

$$a_1^{(m)} = \Theta_{b_1^{(m)}} = 2 \rightarrow \Theta = (1, 3, 4) \rightarrow a_2^{(m)} = \Theta_{b_2^{(m)}} = 4 \rightarrow \Theta = (1, 3)$$
$$\rightarrow a_3^{(m)} = \Theta_{b_3^{(m)}} = 3 \rightarrow \Theta = (1) \rightarrow a_4^{(m)} = \Theta_{b_4^{(m)}} = 1 \rightarrow \Theta = ()$$

Hence, we get $\mathbf{a}^{(m)} = (2, 4, 3, 1)$.

In this scheme, for each block, we first convert the incoming $\lfloor \log_2(N_t!) \rfloor$ bits to an integer m, and then map m to $\mathbf{a}^{(m)}$, dictating the activation order of transmit antennas. At the receiver side, the demapping procedure is straightforward. We can reverse the mapping process and get m from $\mathbf{a}$. Then the integer m is converted to the $\lfloor \log_2(N_t!) \rfloor$ bits.

3.1.2 Average Bit Error Probability Analysis

In this subsection, we examine the performance of the DSM system. For simplicity, we consider $N_t = 2$. We first derive an analytical upper bound on the ABEP for general modulations, and then carry out the asymptotic analysis for the special case of BPSK modulation.

ABEP Upper Bound

Noting that $\Pr(\mathbf{X}_t = [\backslash]) = \Pr(\mathbf{X}_t = [/]) = \frac{1}{2}$, an upper bound on the ABEP can be thus expressed according to the union bounding technique as

$$P_e \leq \sum_{\mathbf{X}_t \in \mathcal{G}, \mathbf{X}_t = [\backslash]} \sum_{\widehat{\mathbf{X}}_t \in \mathcal{G}} \frac{N\left(\mathbf{X}_t \rightarrow \widehat{\mathbf{X}}_t\right) \Pr\left(\mathbf{X}_t \rightarrow \widehat{\mathbf{X}}_t\right)}{M^2\left(1 + 2\log_2(M)\right)}, \tag{3.18}$$

where $[\backslash]$ and $[/]$ indicate a diagonal matrix and an anti-diagonal matrix, respectively, $N(\mathbf{X}_t \rightarrow \widehat{\mathbf{X}}_t)$ measures the number of bits in error between $\mathbf{X}_t$ and $\widehat{\mathbf{X}}_t$, and $\Pr(\mathbf{X}_t \rightarrow \widehat{\mathbf{X}}_t)$ denotes the PEP, which accounts for the probability of the error event that $\widehat{\mathbf{X}}_t$ is decoded given that $\mathbf{X}_t$ is conveyed.

Lemma 1 *The PEPs can be approximately calculated in terms of the Gaussian Q-function as*

$$\begin{aligned} & Pr\left(\mathbf{X}_t \rightarrow \widehat{\mathbf{X}}_t | \mathbf{X}_t = [\backslash], \widehat{\mathbf{X}}_t = [\backslash]\right) \\ & = E_{\mathbf{H}_t}\left\{Q\left(\frac{\Re\{\alpha\}\mathbf{h}_1^H\mathbf{h}_1 + \Re\{\beta\}\mathbf{h}_2^H\mathbf{h}_2}{\sqrt{\frac{1}{\gamma}\left(|\alpha|^2\mathbf{h}_1^H\mathbf{h}_1 + |\beta|^2\mathbf{h}_2^H\mathbf{h}_2\right)}}\right)\right\}, \end{aligned} \tag{3.19}$$

and

$$Pr\left(\mathbf{X}_t \to \widehat{\mathbf{X}}_t | \mathbf{X}_t = [\backslash], \widehat{\mathbf{X}}_t = [/]\right)$$
$$= E_{\mathbf{H}_t}\left\{E_{\mathbf{S}_{t-1}}\left\{Q\left(\sqrt{\frac{\gamma}{2}\left(\begin{array}{c}\mathbf{h}_1^H\mathbf{h}_1 - \Re\left\{\kappa\mathbf{h}_1^H\mathbf{h}_2\right\}\\ +\mathbf{h}_2^H\mathbf{h}_2 - \Re\left\{\mu\mathbf{h}_2^H\mathbf{h}_1\right\}\end{array}\right)}\right)\right\}\right\}, \tag{3.20}$$

respectively, where $\alpha = 1 - x_{2t-1}^*\hat{x}_{2t-1}$, $\kappa = s_{2t-3}^* s_{2t-2} x_{2t-1}^* \hat{x}_{2t-1}$, $\beta = 1 - x_{2t}^*\hat{x}_{2t}$, *and* $\mu = s_{2t-3}s_{2t-2}^* x_{2t}^*\hat{x}_{2t}$. *In (3.19) and (3.20),* $\gamma = 1/N_0$ *is the SNR at each receive antenna,* s_t, x_t, *and* $\hat{x}_t$ *are the transmitted symbol, the information symbol, and the decoded symbol at time slot t, respectively, and* $\mathbf{h}_i = \left[h_{i1}, \ldots, h_{iN_r}\right]^T$ $(i = 1, 2)$ *is the channel vector whose j-th entry denotes the channel gain from the i-th transmit antenna to the j-th receive antenna.*

Proof The PEP can be readily written as

$$\Pr(\mathbf{X}_t \to \widehat{\mathbf{X}}_t) = \Pr\left(\mathrm{Tr}\left\{\Re\left\{\mathbf{Y}_t^H\mathbf{Y}_{t-1}\Delta\mathbf{X}_t\right\}\right\} < 0\right), \tag{3.21}$$

where $\Delta\mathbf{X}_t = \mathbf{X}_t - \widehat{\mathbf{X}}_t$.

First, consider both $\mathbf{X}_t$ and $\widehat{\mathbf{X}}_t$ are diagonal matrices. Therefore, if there exists a decoding error then it must take place in the signal domain. Assuming that $\mathbf{S}_{t-1}$ is a diagonal matrix, it follows that $\mathbf{S}_t$ is also a diagonal matrix and hence

$$\begin{aligned}\mathrm{Tr}\left\{\mathbf{Y}_t^H\mathbf{Y}_{t-1}\Delta\mathbf{X}_t\right\} &= \mathrm{Tr}\left\{\begin{bmatrix}\mathbf{h}_1^H s_{2t-1}^* + \mathbf{n}_{2t-1}^H\\ \mathbf{h}_2^H s_{2t}^* + \mathbf{n}_{2t}^H\end{bmatrix}\begin{bmatrix}\mathbf{h}_1^T s_{2t-3} + \mathbf{n}_{2t-3}^T\\ \mathbf{h}_2^T s_{2t-2} + \mathbf{n}_{2t-2}^T\end{bmatrix}^T\right.\\ &\quad\left.\times\begin{bmatrix}x_{2t-1} - \hat{x}_{2t-1} & 0\\ 0 & x_{2t} - \hat{x}_{2t}\end{bmatrix}\right\}\\ &= \left(\mathbf{h}_1^H s_{2t-1}^* + \mathbf{n}_{2t-1}^H\right)\left(\mathbf{h}_1 s_{2t-3} + \mathbf{n}_{2t-3}\right)\left(x_{2t-1} - \hat{x}_{2t-1}\right)\\ &\quad+\left(\mathbf{h}_2^H s_{2t}^* + \mathbf{n}_{2t}^H\right)\left(\mathbf{h}_2 s_{2t-2} + \mathbf{n}_{2t-2}\right)\left(x_{2t} - \hat{x}_{2t}\right).\end{aligned} \tag{3.22}$$

Recalling the relationship between the transmitted block and the information block, i.e., $s_{2t-3}x_{2t-1} = s_{2t-1}$ and $s_{2t-2}x_{2t} = s_{2t}$, (3.22) can be further simplified as

$$\begin{aligned}\mathrm{Tr}\left\{\mathbf{Y}_t^H\mathbf{Y}_{t-1}\Delta\mathbf{X}_t\right\} =& \alpha\mathbf{h}_1^H\mathbf{h}_1 + \beta\mathbf{h}_2^H\mathbf{h}_2 + s_{2t-3}^*\alpha\mathbf{h}_1^H\mathbf{n}_{2t-3}\\ &+ s_{2t-2}^*\beta\mathbf{h}_2^H\mathbf{n}_{2t-2} + s_{2t-1}\alpha\mathbf{n}_{2t-1}^H\mathbf{h}_1 + s_{2t}\beta\mathbf{n}_{2t}^H\mathbf{h}_2\\ &+ x_{2t-1}\alpha\mathbf{n}_{2t-1}^H\mathbf{n}_{2t-3} + x_{2t}\beta\mathbf{n}_{2t}^H\mathbf{n}_{2t-2}.\end{aligned} \tag{3.23}$$

Therefore, ignoring the second order noise terms in (3.23), when conditioned on the channel gains, $\mathrm{Tr}\left\{\mathbf{Y}_t^H\mathbf{Y}_{t-1}\Delta\mathbf{X}_t\right\}$ is a complex-valued Gaussian distributed RV with mean

$$\lambda_1 = \alpha\mathbf{h}_1^H\mathbf{h}_1 + \beta\mathbf{h}_2^H\mathbf{h}_2, \tag{3.24}$$

and variance

$$\sigma_1^2 = \frac{2}{\gamma}\left(|\alpha|^2 \mathbf{h}_1^H \mathbf{h}_1 + |\beta|^2 \mathbf{h}_2^H \mathbf{h}_2\right). \tag{3.25}$$

For the case in which $\mathbf{S}_{t-1}$ is an anti-diagonal matrix, one can easily find that $\mathrm{Tr}\left\{\mathbf{Y}_t^H \mathbf{Y}_{t-1} \Delta \mathbf{X}_t\right\}$ can be expressed similarly to (3.23) with the only exception being that $\mathbf{h}_1$ and $\mathbf{h}_2$ are switched. Now, (3.19) can be obtained by noticing the facts that the probability that $\mathbf{S}_{t-1}$ is chosen as either a diagonal matrix or an anti-diagonal matrix is equal to one half, and the statistics of $\mathbf{h}_1$ and $\mathbf{h}_2$ are the same.

Second, consider that $\mathbf{X}_t$ is a diagonal matrix while $\widehat{\mathbf{X}}_t$ is an anti-diagonal matrix. This means that the decoding error occurs in the spatial domain. In that case, if $\mathbf{S}_{t-1}$ is a diagonal matrix, we have

$$\begin{aligned}\mathrm{Tr}\left\{\mathbf{Y}_t^H \mathbf{Y}_{t-1} \Delta \mathbf{X}_t\right\} &= \mathbf{h}_1^H \mathbf{h}_1 + \mathbf{h}_2^H \mathbf{h}_2 - \kappa \mathbf{h}_1^H \mathbf{h}_2 - \mu \mathbf{h}_2^H \mathbf{h}_1 + s_{2t} \mathbf{n}_{2t}^H (\mathbf{h}_2 - \mu \mathbf{h}_1) \\ &+ s_{2t-1} \mathbf{n}_{2t-1}^H (\mathbf{h}_1 - \kappa \mathbf{h}_2) + s_{2t-2}^* \left(\mathbf{h}_2^H - \kappa \mathbf{h}_1^H\right) \mathbf{n}_{2t-2} \\ &+ s_{2t-3}^* \left(\mathbf{h}_1^H - \mu \mathbf{h}_2^H\right) \mathbf{n}_{2t-3} + x_{2t-1} \mathbf{n}_{2t-1}^H \mathbf{n}_{2t-3} \\ &- \hat{x}_{2t-1} \mathbf{n}_{2t-1}^H \mathbf{n}_{2t-2} - \hat{x}_{2t} \mathbf{n}_{2t}^H \mathbf{n}_{2t-3} + x_{2t} \mathbf{n}_{2t}^H \mathbf{n}_{2t-2}.\end{aligned} \tag{3.26}$$

Again, excluding the second order noise terms in (3.26), $\mathrm{Tr}\left\{\mathbf{Y}_t^H \mathbf{Y}_{t-1} \Delta \mathbf{X}_t\right\}$ has a complex Gaussian distribution with mean and variance given by

$$\lambda_2 = \mathbf{h}_1^H \mathbf{h}_1 - \kappa \mathbf{h}_1^H \mathbf{h}_2 + \mathbf{h}_2^H \mathbf{h}_2 - \mu \mathbf{h}_2^H \mathbf{h}_1, \tag{3.27}$$

and

$$\sigma_2^2 = \frac{2}{\gamma}\left(\|\mathbf{h}_2 - \kappa \mathbf{h}_1\|^2 + \|\mathbf{h}_1 - \mu \mathbf{h}_2\|^2\right), \tag{3.28}$$

respectively. For the same reason given before, (3.20) can be obtained. □

Now, with the above lemma, the upper bound on the ABEP can be calculated from (3.18). Note that, though (3.18) results from the union bound, the ABEP may be slightly underestimated at very low SNR since the second order noise terms are neglected. However, we will still refer to the result as an upper bound, especially when simulations confirm so for the reasonable SNR range. For $M > 2$, it is difficult to simplify (3.18) as the statistics of the random variables inside the Gaussian Q-function in (3.19) and (3.20) are difficult to characterize. In the sequel, however, we show that a closed-form expression for the upper bound on the ABEP is available for BPSK signaling, i.e., when $M = 2$.

Corollary 1 *In the case of BPSK modulation, the upper bound on the ABEP of the DSM scheme is given by*

$$P_e \leq (1-f(\gamma))^{N_r} \sum_{l=0}^{N_r-1} \frac{1}{2^{N_r-1+l}} \binom{N_r-1+l}{l}$$
$$\times (1+f(\gamma))^l + \sum_{l=0}^{2N_r-1} \frac{1}{3\cdot 2^{2N_r-1+l}} \binom{2N_r-1+l}{l}$$
$$\times \left[(1-f(\gamma))^{2N_r}(1+f(\gamma))^l + 2\left(1-f\left(\frac{\gamma}{2}\right)\right)^{2N_r}\left(1+f\left(\frac{\gamma}{2}\right)\right)^l\right], \quad (3.29)$$

where $f(\gamma) = \sqrt{\gamma/(2+\gamma)}$.

Proof The error events can be explicitly defined as follows:

- E1: $\mathbf{X}_t = [\backslash]$, $\widehat{\mathbf{X}}_t = [\backslash]$, $\widehat{x}_{2t-1} \neq x_{2t-1}$, and $\widehat{x}_{2t} = x_{2t}$;
- E2: $\mathbf{X}_t = [\backslash]$, $\widehat{\mathbf{X}}_t = [\backslash]$, $\widehat{x}_{2t-1} = x_{2t-1}$, and $\hat{x}_{2t} \neq x_{2t}$;
- E3: $\mathbf{X}_t = [\backslash]$, $\widehat{\mathbf{X}}_t = [\backslash]$, $\widehat{x}_{2t-1} \neq x_{2t-1}$, and $\widehat{x}_{2t} \neq x_{2t}$;
- E4: $\mathbf{X}_t = [\backslash]$, $\widehat{\mathbf{X}}_t = [/]$, $\widehat{x}_{2t-1} = x_{2t-1}$, and $\widehat{x}_{2t} = x_{2t}$;
- E5: $\mathbf{X}_t = [\backslash]$, $\widehat{\mathbf{X}}_t = [/]$, $\widehat{x}_{2t-1} \neq x_{2t-1}$, and $\widehat{x}_{2t} = x_{2t}$;
- E6: $\mathbf{X}_t = [\backslash]$, $\widehat{\mathbf{X}}_t = [/]$, $\widehat{x}_{2t-1} = x_{2t-1}$, and $\widehat{x}_{2t} \neq x_{2t}$;
- E7: $\mathbf{X}_t = [\backslash]$, $\widehat{\mathbf{X}}_t = [/]$, $\widehat{x}_{2t-1} \neq x_{2t-1}$, and $\widehat{x}_{2t} \neq x_{2t}$.

For BPSK modulation, symbols are drawn from $\{-1, +1\}$. Consequently, (3.19) can be readily derived as

$$\Pr\left(\mathbf{X}_t \to \widehat{\mathbf{X}}_t | \mathrm{E1}\right) = E_{\mathbf{H}_t}\left\{Q\left(\sqrt{\gamma \mathbf{h}_1^H \mathbf{h}_1}\right)\right\} = \Pr\left(\mathbf{X}_t \to \widehat{\mathbf{X}}_t | \mathrm{E2}\right), \quad (3.30)$$

and

$$\Pr\left(\mathbf{X}_t \to \widehat{\mathbf{X}}_t | \mathrm{E3}\right) = E_{\mathbf{H}_t}\left\{Q\left(\sqrt{\gamma\left(\mathbf{h}_1^H \mathbf{h}_1 + \mathbf{h}_2^H \mathbf{h}_2\right)}\right)\right\}. \quad (3.31)$$

On the other hand, by resorting to the following property:

$$\Pr(s_{2t-3}s_{2t-2} = 1) = \Pr(s_{2t-3}s_{2t-2} = -1) = \frac{1}{2}, \quad (3.32)$$

Equation (3.20) can be simplified as

$$\Pr\left(\mathbf{X}_t \to \widehat{\mathbf{X}}_t | \mathrm{E4}\right) = \Pr\left(\mathbf{X}_t \to \widehat{\mathbf{X}}_t | \mathrm{E7}\right)$$
$$= \frac{1}{2} E_{\mathbf{H}_t}\left\{Q\left(\sqrt{\frac{\gamma}{2}\|\mathbf{h}_1 + \mathbf{h}_2\|^2}\right) + Q\left(\sqrt{\frac{\gamma}{2}\|\mathbf{h}_1 - \mathbf{h}_2\|^2}\right)\right\}, \quad (3.33)$$

and

$$\Pr\left(\mathbf{X}_t \to \widehat{\mathbf{X}}_t|, \mathrm{E5}\right) = E_{\mathbf{H}_t}\left\{Q\left(\sqrt{\frac{\gamma}{2}\left(\mathbf{h}_1^H\mathbf{h}_1 + \mathbf{h}_2^H\mathbf{h}_2\right)}\right)\right\}$$
$$= \Pr\left(\mathbf{X}_t \to \widehat{\mathbf{X}}_t|\mathrm{E6}\right). \quad (3.34)$$

Then, by virtue of [[4], Eqs. (9.5) and (9.6)], (3.30), (3.31), (3.33), and (3.34) can be all solved in closed form. Finally, we have (3.29) from (3.18). □

Asymptotic Analysis

Although (3.29) is expressed as a closed form, it offers little insight into the performance of DSM. To this end, we focus on the derivation of the asymptotic expression of (3.29) when the SNR approaches infinity. According to the power series expansion of $(1+x)^{-1/2}$ [[5], Eq. (1.112.4)], we have

$$P_e \leq \frac{1}{2^{N_r-1}} \sum_{l=0}^{N_r-1} \binom{N_r-1+l}{l} \frac{1}{\gamma^{N_r}} + \text{H.O.T.} \quad (3.35)$$

We see from (3.35) that a full receiver-side diversity order, i.e., N_r, is achieved by a DSM system, which is identical to a coherent SM system using N_r receive antennas, and the coding gain of a DSM system is

$$G_c = 2 \cdot \left[2 \sum_{l=0}^{N_r-1} \binom{N_r-1+l}{l}\right]^{-1/N_r}. \quad (3.36)$$

The coding gain of a coherent SM system can be readily shown to be the same as that of a DSM system, i.e., (3.36), when both systems adopt BPSK modulation.

Note that when DSM and SM both employ BPSK, their rates are different. To equate their rates, DSM needs to use BPSK for one antenna and QPSK for the other. It is well known that QPSK has a 3 dB SNR disadvantage in comparison to BPSK. Hence, DSM with mixed BPSK and QPSK will lead to an SNR loss of less than 3 dB in comparison with the BPSK-SM which has the same spectral efficiency of 2 bps/Hz.

3.1.3 Performance Evaluation

In this subsection, we conduct BER simulations under various system configurations to examine DSM performance. Simulations are carried out over the quasi-static Rayleigh fading channel.

DSM Versus SM

Figure 3.3 compares the BER performance between DSM and SM using $N_t = 4$ transmit antennas. We target a spectral efficiency of 3 bps/Hz. Note that we have used the appropriate modulation orders to obtain the same spectral efficiency, i.e., QPSK

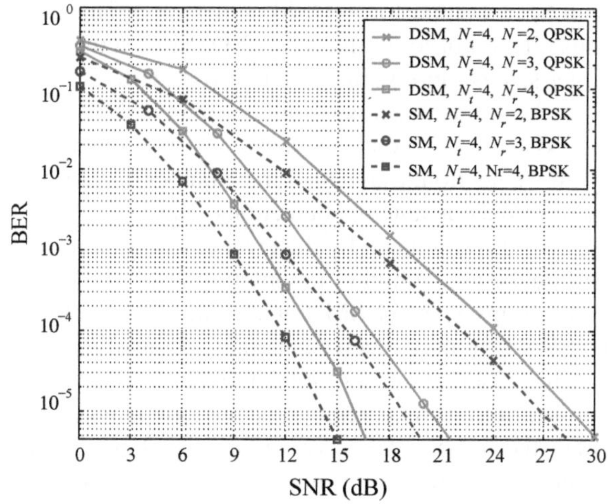

Fig. 3.3 BER performance of differential detection DSM versus coherent detection SM at 3 bps/Hz transmission rate with $N_t = 4$, where DSM uses QPSK whereas SM uses BPSK

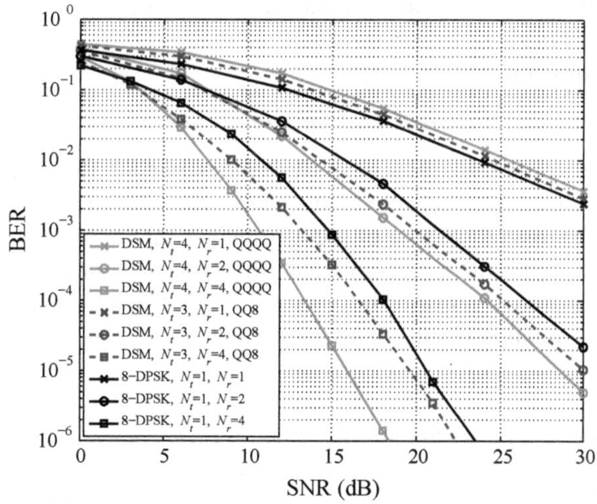

Fig. 3.4 BER performance of differential detection DSM with $N_t = 3$ and $N_t = 4$ at 3 bps/Hz transmission rate. DSM with $N_t = 3$ uses QPSK, QPSK and 8PSK for the three time instants in each block, whereas DSM with $N_t = 4$ always uses QPSK. "Q" and "8" in the legend stand for QPSK and 8PSK, respectively

for DSM and BPSK for SM. It can be observed that the loss in BER performance induced by the differential scheme is no more than 3 dB. Simulation results show that though DSM uses higher modulation order than SM in order to make up for its spectral efficiency loss in the spatial domain, a 3 dB degradation is still retained.

DSM with Multiple Transmit Antennas

Figure 3.4 compares the DSM BER performance with multiple transmit antennas. The transmission rate is fixed at 3 bps/Hz. For $N_t = 4$, DSM always uses QPSK and for $N_t = 3$, DSM uses QPSK, QPSK, and 8PSK for the three time instants in each block. The performance of a single-antenna system adopting 8-DPSK is also considered as a benchmark. It can be observed that in the cases of $N_r = 2, 4$, DSM with $N_t = 4$ outperforms DSM with $N_t = 3$. This can be accounted for a higher signal constellation order in the $N_t = 3$ DSM than the $N_t = 4$ DSM. The constellation of larger size is more vulnerable to errors. However, in the case of $N_r = 1$, DSM with $N_t = 3$ exhibits better performance than DSM with $N_t = 4$. This can be explained by the existence of two error types: (1) erroneous antenna detection and (2) erroneous symbol detection. Using the same number of receive antennas, DSM with $N_t = 4$ has more erroneous antenna detections than DSM with $N_t = 3$. In the case of $N_r = 1$, the error is dominated by the erroneous estimation of transmit antennas. Therefore, despite the use of higher modulation order in DSM with $N_t = 3$, $N_t = 3$ DSM still outperforms $N_t = 4$ DSM. For a similar reason, DSM outperforms the single-antenna DPSK when more receive antennas are involved.

This observation is significant as it reveals two phenomena. The first is that the number of transmit antennas and the size of the signal constellation can be traded off. When the number of receive antennas is relatively small, fewer transmit antennas with higher modulation order may be preferable. When the number of receive antennas is sufficient, it may be better to use more transmit antennas with lower modulation order. The second is that the number of receive antennas is crucial to the performance of DSM. To obtain the full potential of DSM, the system should be equipped with sufficient receive antennas.

DSM Versus Other Transmission Schemes

In Fig. 3.5, for reference, we demonstrate DSM's performance versus three different transmission schemes with fixed transmission rate at 3 bps/Hz. The first one is the differential scheme of the well known Alamouti (DA) scheme with $N_t = 2$ [1]. The second is the single-antenna differential scheme, DPSK. The third is the differential scheme of quasi-orthogonal space-time block code (DQOSTBC) [6] with $N_t = 4$. Simulation results show that when $N_r = 1$, DSM exhibits the worst performance.

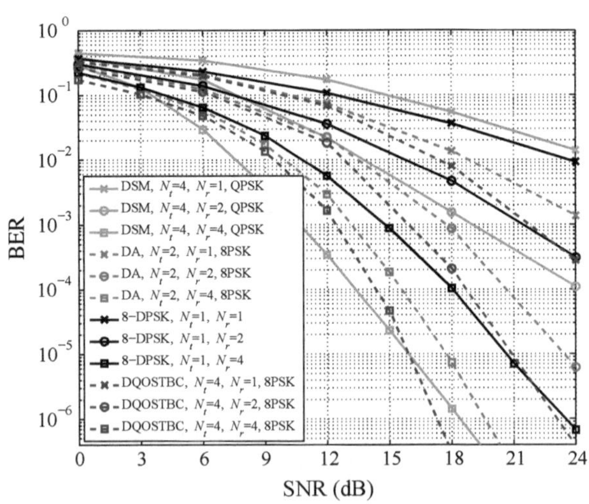

Fig. 3.5 BER performance of DSM with $N_t = 4$ versus DA, 8-DPSK and DQOSTBC at 3 bps/Hz transmission rate. DSM with $N_t = 4$ uses QPSK and the others use 8PSK

However, when N_r increases to 4, DSM outperforms DA, single antenna DPSK and DQOSTBC. This is because when $N_r = 1, 2$, the erroneous antenna detection of DSM is so significant that it offsets its coding gain. The increasing of N_r reduces the erroneous antenna detection of DSM considerably. Therefore, when $N_r = 4$ the coding gain of DSM due to the use of lower-order modulation dominates the result and thus DSM outperforms the others.

In addition, it can be observed that in all antenna configurations, DA and DQOSTBC benefit more from increasing the SNR than does DSM. This is because the diversity order of DSM approaches that of single antenna system which is N_r as detailed in [7]. In contrast, Alamouti with $N_t = 2$ provides full diversity, i.e., $2N_r$ and QOSTBC with $N_t = 4$ provides half of the maximum achievable diversity. Thus, at high SNR, DA and DQOSTBC outperform DSM.

3.2 Design Guidelines for Differential Spatial Modulation

In this section, we first design a low-complexity near-ML receiver. Then, we present the idea of Gray coding for improving the coding gain and diversity gain achieved by DSM system.

3.2.1 Low-Complexity Near-ML Receiver

As indicated in (3.15), the ML detector has to identify the most probable order out of nearly $\log_2(N_t!)$ antenna activation orders from the received signals within N_t time instants, leading to a very high search complexity. In this subsection, we propose a low-complexity near-ML receiver to solve this problem. To begin with, let $\mathbf{W}_t = \mathbf{Y}_t^H \mathbf{Y}_{t-1}$, which is of dimension $N_t \times N_t$. Due to the independence among the modulated symbols transmitted in different time slots, (3.15) can be simplified as

$$\hat{L}_t(\tilde{m}, i) = \arg\max_{\tilde{s}_i \in \chi_i} \Re\left\{W_t\left(i, a_i^{(\tilde{m})}\right)\tilde{s}_i\right\}, \tag{3.37}$$

and

$$\begin{cases} \hat{m} = \arg\max\limits_{\tilde{m}\in\omega} \sum\limits_{i=1}^{N_t} \Re\left\{W_t\left(i, a_i^{(\tilde{m})}\right)\hat{L}_t(\tilde{m}, i)\right\} \\ \hat{s}_n = \hat{L}_t\left(\hat{m}, n\right), \ n = 1, \ldots, N_t \end{cases}, \tag{3.38}$$

where $\hat{\mathbf{L}}_t$ is of dimension $2^{\lfloor \log_2(N_t!)\rfloor} \times N_t$ and $\omega = \{1, \ldots, 2^{\lfloor \log_2(N_t!)\rfloor}\}$. Here, to be more general, we assume that the i-th modulated symbol is drawn from χ_i. Since given $\tilde{m}_1 \in \omega$, there always exists at least an $\tilde{m}_2 \in \omega$ with $\tilde{m}_2 \neq \tilde{m}_1$ that satisfies $a_i^{(\tilde{m}_2)} = a_i^{(\tilde{m}_1)}$ for some $i \in \xi(= \{1, \ldots, N_t\})$ and large N_t, one will discover from (3.37) that $L_t(\tilde{m}_2, i) = L_t(\tilde{m}_1, i)$ in those cases. To avoid duplicate computation, we can further simplify (3.37) and (3.38) as

$$\hat{G}_t(i,j) = \arg\max_{\tilde{s}_i \in \chi_i} \Re\{W_t(i,j)\,\tilde{s}_i\}, \tag{3.39}$$

and

$$\begin{cases} \hat{m} = \arg\max_{\tilde{m}\in\omega} \sum_{i=1}^{N_t} D_t\left(i, a_i^{(\tilde{m})}\right) \\ \hat{s}_n = \hat{G}_t\left(n, a_n^{(\hat{m})}\right), \ n = 1, \ldots, N_t \end{cases}, \tag{3.40}$$

where $D_t(i,j) = \Re\left\{W_t(i,j)\,\hat{G}_t(i,j)\right\}$ with $i,j \in \xi$ and both $\hat{\mathbf{G}}_t$ and $\mathbf{D}_t$ are of dimension $N_t \times N_t$. The proposed detector is based on (3.39) and (3.40), which is carried out in the following two steps. The first step is to solve (3.39), and the second step is to solve (3.40) but with $\tilde{m} \in \omega$ relaxed to $\tilde{m} \in \{1, \ldots, (N_t-1)(N_t-1)!\}$ so as to facilitate low-complexity detection, which will be discussed later. Note that the above-mentioned relaxation only applies to $N_t \geq 3$ cases, which are of particular interest as we focus on the reduction of the computational complexity for a large N_t. With such relaxation, however, the detector may decide on $\hat{m}$ that is not included in ω, thus resulting in performance loss. Nevertheless, we will see from the simulation results that this performance loss is negligible. In the sequel, details on how each step works will be presented, and we will omit the notation t for brevity.

First Step: Signal Domain Pre-detection

The first step aims at low-complexity calculation of the matrix $\hat{\mathbf{G}}$ defined in (3.39). This step invokes an initial detection of modulated symbols and thus is termed signal domain pre-detection. The idea is clarified in the following.

Let us reformulate (3.39) as

$$\hat{G}(i,j) = \arg\max_{\tilde{s}_i \in \chi_i} \{\Re\{W(i,j)\}\Re\{\tilde{s}_i\} - \Im\{W(i,j)\}\Im\{\tilde{s}_i\}\}. \tag{3.41}$$

The above equation indicates that $\hat{G}(i,j)$ is a modulated symbol drawn from χ_i, for which the maximum of the expression is achieved. From (3.41), it can be deduced that $\Re\{\hat{G}(i,j)\}$ must have the same polarity as $\Re\{W(i,j)\}$ and $\Im\{\hat{G}(i,j)\}$ must have the opposite polarity to $\Im\{W(i,j)\}$. Therefore, we can solve (3.41) by first searching for the constellation point, $\breve{s}_i$, belonging to the first quadrant, which satisfies

$$\begin{aligned} D(i,j) &= \Re\{W(i,j)\}\Re\left\{\hat{G}(i,j)\right\} - \Im\{W(i,j)\}\Im\left\{\hat{G}(i,j)\right\} \\ &= |\Re\{W(i,j)\}|\,\Re\left\{\breve{s}_i\right\} + |\Im\{W(i,j)\}|\,\Im\left\{\breve{s}_i\right\}, \end{aligned} \tag{3.42}$$

based on the symmetry of the PSK constellation and then mapping $\breve{s}_i$ to $\hat{G}(i,j)$ based on the above-mentioned polarity property in analogy with [8]. However, we find that a further reduction of search complexity, which accounts for a half fewer constellation points than [8], is possible, thanks to the symmetry of the real and imaginary parts of the constellation points belonging to the first quadrant. To show this, define

$$\Delta(i,j) = \begin{cases} 0, & |\Re\{W(i,j)\}| > |\Im\{W(i,j)\}| \\ 1, & \text{otherwise} \end{cases} \tag{3.43}$$

Further, let $\hat{G}(i,j) = e^{\sqrt{-1}\frac{2\pi}{M_i}K(i,j)}$ and $\breve{s}_i = e^{\sqrt{-1}\frac{2\pi}{M_i}\left(K'(i,j)+\frac{M_i}{8}\Delta(i,j)\right)}$. By definition, the solutions of $\breve{s}_i$ and $\hat{G}(i,j)$ are now translated into the solutions of $K'(i,j)$ and $K(i,j)$, respectively. Since given $a_1 > a_2 > 0$ and $b_1 > b_2 > 0$ it follows that $a_1b_1 + a_2b_2 > a_1b_2 + a_2b_1$, from (3.42) the search for $\breve{s}_i$ can be completed by

$$\begin{aligned} K'(i,j) = \underset{k\in\left\{0,\ldots,\frac{M_i}{8}\right\}}{\arg\max} & \left\{ |\Re\{W(i,j)\}| \cos\left(\frac{2\pi k}{M_i} + \frac{\pi\Delta(i,j)}{4}\right) \right. \\ & \left. + |\Im\{W(i,j)\}| \sin\left(\frac{2\pi k}{M_i} + \frac{\pi\Delta(i,j)}{4}\right) \right\}. \end{aligned} \tag{3.44}$$

Then, according to $\hat{G}(i,j) = \text{sign}\{\Re\{W(i,j)\}\} \cdot \Re\{\breve{s}_i\} - \sqrt{-1}\,\text{sign}\{\Im\{W(i,j)\}\} \cdot \Im\{\breve{s}_i\}$, the relationship between $K'(i,j)$ and $K(i,j)$ can be readily established as

$$\begin{aligned} K(i,j) = & \frac{M_i}{4}\left(2 + \text{sign}\{\Im\{W(i,j)\}\} + \frac{\text{sign}\{\Re\{W(i,j)\}\}}{\text{sign}\{\Im\{W(i,j)\}\}}\right) \\ & - \frac{\text{sign}\{\Re\{W(i,j)\}\}}{\text{sign}\{\Im\{W(i,j)\}\}}\left(K'(i,j) + \frac{M_i}{8}\Delta(i,j)\right). \end{aligned} \tag{3.45}$$

Example Assume 16-PSK modulation for the i-th time slot, i.e., $M_i = 16$ and $\chi_i = \left\{1, e^{2\sqrt{-1}\pi/16}, \ldots, e^{2\sqrt{-1}\pi 15/16}\right\}$. Suppose that $W(i,j) = -2 - 3\sqrt{-1}$. Thus, we have $\Delta(i,j) = 1$, $\text{sign}\{\Re\{W(i,j)\}\} = -1$, and $\text{sign}\{\Im\{W(i,j)\}\} = -1$. Searching through about one-eighth of the constellation points according to (3.44) gives $K'(i,j) = 1$ and accordingly $\breve{s}_i = e^{2\sqrt{-1}\pi 3/16}$. Then, from (3.45) we have $K(i,j) = 5$ and accordingly $\hat{G}(i,j) = e^{2\sqrt{-1}\pi 5/16}$, which agrees with that solved by (3.41) via searching through the whole constellation (Fig. 3.6).

When $\hat{\mathbf{G}}$ is ready, it is straightforward to obtain the metric matrix $\mathbf{D}$ from (3.42).

Second Step: Joint Detection

As discussed earlier, the second step is to search for an optimal order of indices of transmit antennas, parametrized by $\hat{m}$, via

$$\hat{m} = \underset{\tilde{m}\in\{1,\ldots,(N_T-1)(N_t-1)!\}}{\arg\max} \sum_{i=1}^{N_t} D\left(i, a_i^{(\tilde{m})}\right), \tag{3.46}$$

and to search for the most-likely transmitted symbols through the candidates included in $\hat{\mathbf{G}}$. This step invokes both signal domain and spatial domain detection and thus is termed joint detection in this letter. As the signal domain detection is simply $\hat{s}_n = \hat{G}\left(n, a_n^{(\hat{m})}\right)$ with $n = 1, \ldots, N_t$, we will mainly focus on the spatial domain detection in the following.

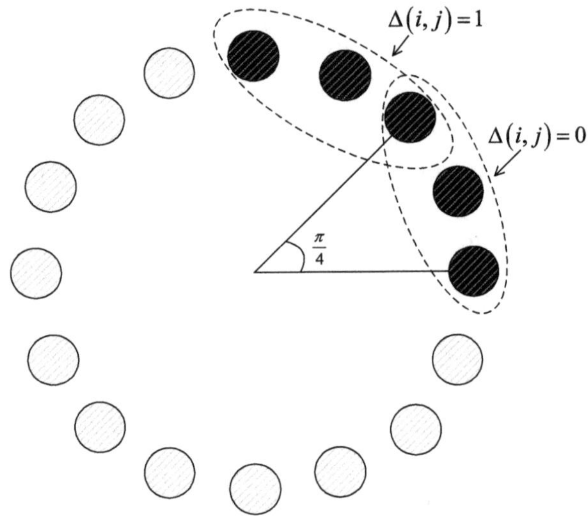

Fig. 3.6 Search zones corresponding to $\Delta(i,j)=0$ and $\Delta(i,j)=1$, respectively, for 16-PSK modulation

As indicated in (3.46), to perform the spatial domain detection, we have to first initialize $\{a^{(1)},\ldots,a^{((N_t-1)(N_t-1)!)}\}$, which becomes impractical for a larger N_t. Noticing that the metric accumulation in (3.46) is dictated by the sequence $a^{(\tilde{m})}$, which has memory, we are motivated by the Viterbi algorithm [9] to reduce the search complexity. The idea can be clarified through the following example. Assume that $N_t=4$ and there exist two candidates $\tilde{m}_1$ and $\tilde{m}_2$, giving $a^{(\tilde{m}_1)}=\{1,2,3,4\}$ and $a^{(\tilde{m}_2)}=\{2,1,3,4\}$. Then, from (3.46), it can be inferred that $\tilde{m}_1$ is preferred over $\tilde{m}_2$ when $D(1,1)+D(2,2)+D(3,3)+D(4,4)>D(1,2)+D(2,1)+D(3,3)+D(4,4)$, and vice versa. In fact, the above decision can be made merely via comparing the following two accumulated metrics $D(1,1)+D(2,2)$ and $D(1,2)+D(2,1)$. Moreover, to extend, provided that $D(1,1)+D(2,2)$ is larger (smaller) than $D(1,2)+D(2,1)$, the antenna activation orders with the first two indices being $\{2,1\}$ ($\{1,2\}$) can be all removed from the search space.

To realize this idea, we find it useful to predefine some parameters, as listed in Table 3.2, and introduce the following enumeration method. The adopted enumeration method is based on the combinational number system (CNS) [3], which enables us to map a natural number $r\in\{1,\ldots,\mathrm{C}(N_t,n)\}$ to a unique permutation $\{c_n,\ldots,c_1\}$ with $c_1,\ldots,c_n\in\{1,\ldots,N_t\}$ and $c_n>\cdots>c_1$, via

$$r=\mathrm{C}(c_n-1,n)+\cdots+\mathrm{C}(c_1-1,1)+1. \tag{3.47}$$

The enumeration procedure starts by choosing the maximal c_n that satisfies $\mathrm{C}(c_n-1,n)\le r-1$, and proceeds by choosing the maximal c_{n-1} that satisfies $\mathrm{C}(c_{n-1}-1,n-1)\le r-\mathrm{C}(c_n-1,n)-1$ and so on until (3.47) is satisfied. As an example, for $N_t=5$, $n=3$, and $r=6$, the permutation $\{c_3,c_2,c_1\}$ can be calculated as $\{5,3,1\}$. With the above-mentioned preliminaries, we summarize the entire

Table 3.2 Definition of parameters in Algorithm 1

ψ^{pre}	a set that stores all most-likely antenna activation orders up to the last time slot
ψ^{now}	a set that stores all most-likely antenna activation orders up to the current time slot
$\mathbf{d}^{pre}$	a vector that stores accumulated metrics corresponding to those antenna activation orders in ψ^{pre}
$\mathbf{d}^{now}$	a vector that stores accumulated metrics corresponding to those antenna activation orders in ψ^{now}
ζ	a set that stores the indices of transmit antennas that are probably chosen at the next time slot

Algorithm 1: Viterbi decoding based spatial domain detector

Initialization: $\psi^{pre} := \xi, \mathbf{d}^{pre} := D(1,:), d^{pre}(N_t) := -\infty$
for $i = 2 : N_t$ *% loop 1 starts*
  $\mathbf{d}^{now} := [0, \ldots, 0]$ % $\mathrm{C}(N_t, i-1)$ zeros
  for $z = 1 : \mathrm{C}(N_t, i-1)$ *% loop 2 starts*
    $\zeta := \xi \backslash \psi^{pre}\{z\}$
    for $k = 1 : N_t - i + 1$ *% loop 3 starts*
      Sort $\psi^{pre}\{z\} \oplus \zeta\{k\}$ in descending order
      and get its address via (3.47) as r
      if $d^{pre}(z) + D(i, \zeta\{k\}) > d^{now}(r)$
        $\psi^{now}\{r\} := \psi^{pre}\{z\} \oplus \zeta\{k\}$
        $d^{now}(r) := d^{pre}(z) + D(i, \zeta\{k\})$
      end if
    end for *% loop 3 ends*
  end for *% loop 2 ends*
  $\mathbf{d}^{pre} := \mathbf{d}^{now}, \psi^{pre} := \psi^{now}$
end for *% loop 1 ends*
Transfer ψ^{now} to the corresponding parameter $\hat{m}$

process of the proposed spatial domain detection in Algorithm 1. In the algorithm, the initialization corresponds to the 1-st time slot case, where $d^{pre}(N_t) := -\infty$ guarantees that $\tilde{m} \in \{1, \ldots, (N_t - 1)(N_t - 1)!\}$ This implies that all antenna activation orders starting with index N_t are excluded in the search. Note that it is also feasible to enable $\tilde{m} \in \omega$ in order to achieve the optimal performance, which, however, will significantly complicate the search as we have to judge the legality of each possible antenna activation order up to the current time slot. For the $(i \geq 2)$-th time slot case, in total $(N_t - i + 1)\,\mathrm{C}(N_t, i-1)$ permutations of i out of N_t transmit antenna indices are to be grouped into $\mathrm{C}(N_t, i)$ classes. In each class, i permutations of the same i indices will compete with each other and only the one with the largest accumulated metric will be saved and fed to the $(i + 1)$-th time slot. The process continues until $i = N_t$, i.e., all transmit antennas have been considered in the comparison.

Table 3.3 Parameters evolution in the example

	$i = 1$	$i = 2$	$i = 3$
ψ^{pre}	$\{1, \ldots, N_T\}$	$\{1, \ldots, N_T\}$	$\{\{1, 2\}, \{1, 3\}, \{2, 3\}\}$
ψ^{now}	–	$\{\{1, 2\}, \{1, 3\}, \{2, 3\}\}$	$\{1, 2, 3\}$
$\mathbf{d}^{pre}$	$[1, 0.5, -\infty]$	$[1, 0.5, -\infty]$	$[1.6, 1.8, 1.3]$
$\mathbf{d}^{now}$	–	$[1.6, 1.8, 1.3]$	2.5

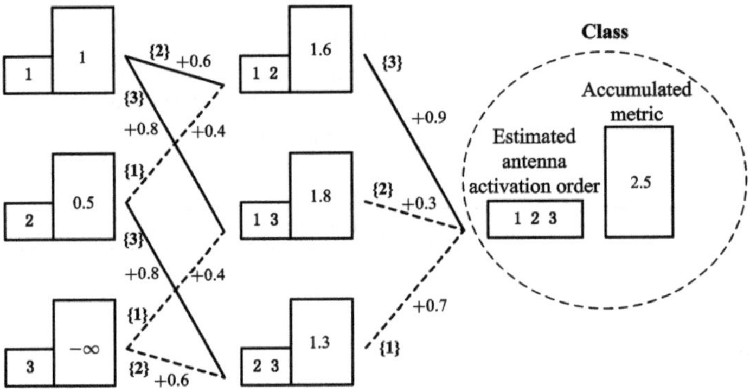

Fig. 3.7 Illustration of the example, where the *solid line* represents the survivor path whereas the *dotted line* represents the discarded path

Example Suppose that $N_T = 3$ and

$$\mathbf{D} = \begin{bmatrix} 1 & 0.5 & 0.2 \\ 0.4 & 0.6 & 0.8 \\ 0.7 & 0.3 & 0.9 \end{bmatrix}.$$

According to Algorithm 1, the evolution of some important parameters for different time slots (before line 14) is presented in Table 3.3. A graphical illustration is also provided in Fig. 3.7.

3.2.2 *Gray Coding for Antenna Activation Orders*

In DSM, the antenna activation orders are used as an information carrying mechanism. Consequently, the difference between antenna activation orders plays an important role in the BER performance of DSM. In previous study, the antenna activation orders are set to follow the lexicographic order such that similar permutations may lead to huge bit difference between the corresponding information bit sequences. This, however, will degrade the BER performance of DSM as a detection error most probably occurs between similar permutations at high SNR. To enhance the DSM

Table 3.4 A look-up table for $N_t = 3$

Bits in lexicographic order	Permutations	Bits in gray code order	Permutations
00	(1 2 3)	00	(1 2 3)
01	(1 3 2)	01	(1 3 2)
10	(2 1 3)	11	(3 1 2)
11	(2 3 1)	10	(3 2 1)

performance, in this subsection we present the idea of encoding permutations of antenna indices in the gray code order for DSM.

Look-Up Table

Table 3.4 shows the mapping between the information bits and the corresponding permutations in the lexicographic order and gray code order for $N_t = 3$, respectively, where two out of in total six permutations have been discarded. Note that although not specified, it can be readily figured out that both manners give rise to the same result for $N_t = 2$. From Table 3.4, it is obvious that the discarded permutations in the gray code order are different from those in the lexicographic order. However, one can see surprisingly that the lexicographic order for $N_t = 3$ also satisfies the requirement of the gray code order, i.e., any two permutations having only two indices difference differ from only one bit. This implies that both manners for $N_t = 3$ will result in the same BER performance. For $N_t > 3$, the lexicographic order cannot meet the above requirement any longer and both manners will lead to totally different performance. This will be verified in the sequel.

Ranking and Unranking Methods

The look-up table method necessitates a storage of all permutations at both the transmitter and receiver, which becomes impractical for large N_t. For example, 36,28,800 and 2.4329×10^{18} permutations need to be stored for $N_t = 10$ and 20, respectively. Obviously, it is impractical to store them for current computers. Therefore, it is advisable to create an easy-to-implement one-to-one mapping from the information bits to the corresponding permutations, called unranking, and inverse mapping from the permutations to the corresponding information bits, called ranking.

To begin with, we have to first figure out how to generate permutations in the gray code order. Aiming at this, we resort to the idea presented in [10]. Table 3.5 gives a solution for $N_t = 4$. From this example, we can expect some important properties of gray code order for a general N_t as follows. For ease of exposition, let us define left and right moving directions for number N_t as $\overleftarrow{N_t}$ and $\overrightarrow{N_t}$, respectively. Also, define a directional indicator for number N_t as $\mathcal{I}_{N_t}$, where $\mathcal{I}_{N_t} = 1$ stands for the situation of $\overleftarrow{N_t}$ and $\mathcal{I}_{N_t} = 0$ for the situation of $\overrightarrow{N_t}$. If the whole swaps between number N_t and the adjacent element in the set $\{1, 2, \ldots, N_t\}$ for the same direction is said to be one *round* for number N_t, one will see that the total number of rounds of number N_t, denoted by L_R, is $(N_t - 1)!$. In addition, it can be concluded that $\mathcal{I}_{N_t} = 1$ if l_{N_t} is odd and $\mathcal{I}_{N_t} = 0$ if l_{N_t} is even, where l_{N_t} is a variable indicating that number N_t is in

Table 3.5 Generating permutations of $\{1, 2, 3, 4\}$ in the gray code order

1234,	1243,	1423,	4123,	$\overleftarrow{4}$
4132,	1432,	1342,	1324,	$\overrightarrow{4}$
3124,	3142,	3412,	4312,	$\overleftarrow{4}$
4321,	3421,	3241,	3214,	$\overrightarrow{4}$
2314,	2341,	2431,	4231,	$\overleftarrow{4}$
4213,	2413,	2143,	2134.	$\overrightarrow{4}$

the l_{N_t}-th round. Details on the generation rule can be referred to the Trotter-Johnson ranking and unranking algorithms [10]. In what follows, we summarize them in our notations and mainly focus on their application to DSM.

Assume that we have a permutation $\mathbf{a} = \{a_1, a_2, \ldots, a_{N_t}\}$ where $u, a_u \in \{1, 2, \ldots, N_t\}$. Define P_{a_u} as the position of the element a_u in $\mathbf{a}^{[a_u]}$, where $\mathbf{a}^{[a_u]}$ indicates the sub-permutation of $\mathbf{a}$, which discards the elements greater than a_u. Initially, we will determine whether $\mathbf{a}$ is generated from the permutations $\{1, 2\}$ or $\{2, 1\}$ for obtaining l_3. It is obvious that $\mathbf{a}$ is generated from $\{1, 2\}$ if $P_2 = 2$, which indicates $l_3 = 1$ ($\mathcal{I}_3 = 1$), while $\mathbf{a}$ is generated from $\{2, 1\}$ if $P_2 = 1$, which indicates $l_3 = 2$ ($\mathcal{I}_3 = 0$). For the case of $N_t > 3$, $\{l_j\}_{j=4}^{N_t}$ can be calculated recursively by

$$l_j = (l_{j-1} - 1) \cdot (j - 1) + \bar{\mathcal{I}}_{j-1} \cdot P_{j-1} + \mathcal{I}_{j-1} \cdot (j - P_{j-1}), \tag{3.48}$$

where $\bar{\mathcal{I}}_{j-1} = 1 - \mathcal{I}_{j-1}$. Finally, the integer $d \in \{1, \ldots, 2^{m_1}\}$ corresponding to $\mathbf{a}$ is calculated by

$$d = (l_{N_t} - 1) \cdot N_t + \bar{\mathcal{I}}_{N_t} \cdot P_{N_t} + \mathcal{I}_{N_t} \cdot (N_t + 1 - P_{N_t}). \tag{3.49}$$

Then, the information bit sequence in the gray code order can be directly obtained from d.

The unranking process is the inversion of the ranking process. Firstly, the information bit sequence in the gray code order is converted into an integer d. Then, l_{N_t} and P_{N_t} can be derived as $l_{N_t} = \lceil d/N_t \rceil$ and $P_{N_t} = < d >_{N_t} + 1$, respectively. For the case of $N_t > 3$, l_k and P_k with $k = N_t - 1$ can be calculated by $l_k = \lceil l_{k+1}/k \rceil$ and $P_k = < l_{k+1} >_k + 1$, respectively. Then, we set $k := k - 1$ and iteratively obtain l_k and P_k, respectively. The process continues until $k = 3$, when $\{l_k\}_{k=3}^{N_t}$ and $\{P_k\}_{k=3}^{N_t}$ are all ready. Note that $\{\mathcal{I}_k\}_{k=3}^{N_t}$ can be determined by $\{l_k\}_{k=3}^{N_t}$. Finally, we can rebuild the permutation from the ranking algorithm.

SNR Gain Analysis

We now analyze the performance of the gray code order by comparing it with that of the lexicographic order. To isolate the effect of Gray coding, we consider the special case of DSM, i.e., differential space shift keying (DSSK), in which $\{s_i\}_{i=1}^{N_t}$ are set to 1s for all time.

An upper bound on the ABEP can be derived according to the union bounding technique as

$$P_e \leq \frac{1}{m_1 \cdot 2^{m_1}} \sum_{p=1}^{2^{m_1}} \sum_{q=1}^{2^{m_1}} N(\mathbf{X}_p \to \widehat{\mathbf{X}}_q) \Pr(\mathbf{X}_p \to \widehat{\mathbf{X}}_q), \tag{3.50}$$

where $\Pr(\mathbf{X}_p \to \widehat{\mathbf{X}}_q)$ is the pairwise error probability accounting for the probability of detecting the information matrix $\widehat{\mathbf{X}}_q$ when $\mathbf{X}_p$ is transmitted, and $N(\mathbf{X}_p \to \widehat{\mathbf{X}}_q)$ is the number of bits in difference between $\mathbf{X}_p$ and $\widehat{\mathbf{X}}_q$. Assuming a rich scattering environment, at high SNR the upper bound in (3.50) can be further approximated as

$$P_e \leq \frac{c \cdot \mathrm{SNR}^{-r_{min} N_R}}{m_1 \cdot 2^{m_1}} \sum_{p,q} N(\mathbf{X}_p \to \widehat{\mathbf{X}}_q | R_{p,q} = r_{min}), \tag{3.51}$$

where c is a constant, $N(\cdot|\cdot)$ is the conditional number of bits in error, $R_{p,q} = \mathrm{rank}\{\mathbf{X}_p - \widehat{\mathbf{X}}_q\}$ and $r_{min} = \min R_{p,q}$. It can be readily figured out that for both the gray code order and lexicographic order we have $r_{min} = 1$, which implies that DSSK achieves unit transmit diversity order regardless of which encoding manner is employed. Therefore, the performance of the gray code order and lexicographic order differ from the SNR gain only. To evaluate the value, let us define the total number of bits in difference with $R_{p,q} = 1$ for both manners as $N^G_{error} = \sum_{p,q} N^G(\mathbf{X}_p \to \widehat{\mathbf{X}}_q | R_{p,q} = 1)$ and $N^L_{error} = \sum_{p,q} N^L(\mathbf{X}_p \to \widehat{\mathbf{X}}_q | R_{p,q} = 1)$, where the subscripts G and L refer to the gray code order and lexicographic order, respectively. Then, from (3.51) the SNR gain achieved by the gray code order over lexicographic order in dB, which is defined based on the BER criterion, can be thus calculated by

$$G_{SNR} = \frac{10}{N_r} \log_{10}(N^L_{error} / N^G_{error}). \tag{3.52}$$

Figure 3.8 shows the comparison results between the BER performances of the gray code order and lexicographic order for $N_t = 4, 6$ and $N_r = 1, 2, 3, 4$ in DSSK under the spectral efficiencies of 1 and 1.5 bps/Hz, respectively. At BER $= 10^{-3}$, it can be seen that in the case of $N_r = 1$ the proposed gray code order achieves about 1.2 and 0.3 dB SNR gains over the lexicographic order for $N_t = 4$ and $N_t = 6$, respectively. Note that the performance gain in DSM is still impressive though it becomes a little smaller than that in DSSK. In addition, from the figure, it is notable that the SNR gain brought by coding decrease as N_r increases for a given N_t.

Figure 3.9 shows the calculated SNR gains achieved by the gray code order over the lexicographic order for $N_t = \{3, 4, 5, 6\}$ and $N_r = 1$ from (3.52) in DSSK. It is expected that no SNR gain is available for $N_t = 3$ and it achieves almost 1.2 dB SNR gain for $N_t = 4$, in which $N^L_{error} = 136$ and $N^G_{error} = 104$. On the other hand, we see that the theoretical results match their simulation counterparts in Fig. 3.8.

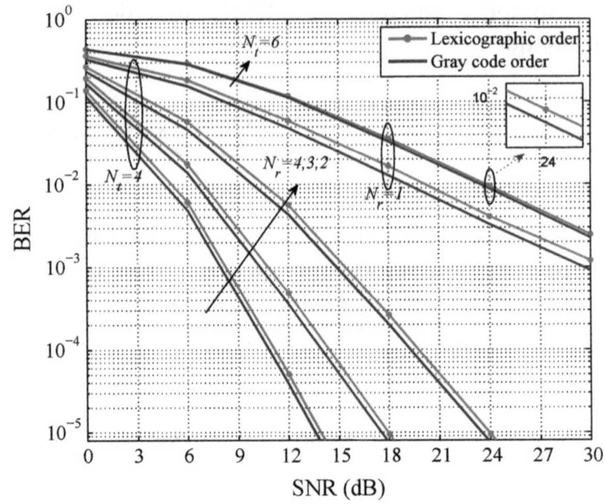

Fig. 3.8 BER performance between the gray code order and lexicographic order for different configurations in DSSK

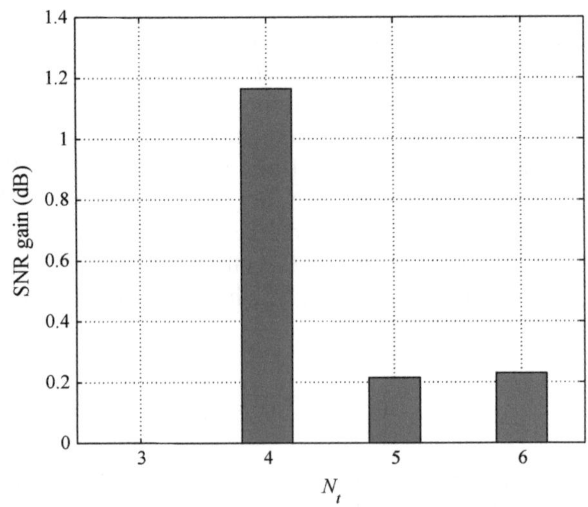

Fig. 3.9 The coding gain ratio between the gray code order and lexicographic order for $N_t = \{3, 4, 5, 6\}$ and $N_r = 1$ in DSSK

3.2.3 Diversity Enhancing Scheme

From the analysis in Sect. 3.2.2, we see that the transmit diversity order achieved by DSM systems depends on r_{min}. Since in the gray code order two adjacent permutations have two elements difference, the minimum rank equals one and in return the transmit diversity order remains unit. To improve the diversity performance of DSM, in this subsection we propose a novel scheme, i.e., intersect (I-)gray code order, based on the gray code order.

In the I-gray code order, the new permutations consist of the ones which locate the odd (or even) positions of the full permutations in the gray code order. The example

Table 3.6 I-gray code order for $N_t = 3$

Candidates in gray code order	Permutations in gray code order	Selected permutations	Permutations in I-gray code order
(1 2 3)			
(1 3 2)	(1 2 3)	**(1 2 3)**	**(1 2 3)**
(3 1 2)	(1 3 2)	**(3 1 2)**	**(3 1 2)**
(3 2 1)	(3 1 2)	**(2 3 1)**	
(2 3 1)	(3 2 1)		
(2 1 3)			

for $N_t = 3$ is given in Table 3.6, where only the odd positions, i.e., 1st, 3rd and 5th, of the full permutations in the gray code order are selected for the purpose. To modulate the information bits, however, only $2^{\lfloor \log_2(N_t!/2) \rfloor} = 2$ permutations are permitted to be used, which obtains the legitimate permutations as (1 2 3) and (3 1 2). It is clear that two adjacent permutations in the I-gray code order will have three elements difference, which improves an additional transmit diversity order in DSSK.

The direct application of the I-gray code order to conventional DSM fails to achieve a transmit diversity order of two since the independent symbol-by-symbol transmission scheme will limit r_{min} to be unit. To overcome this problem, it is advisable to introduce correlation between two adjacent modulated symbols in a DSM block under the framework of the I-gray code order. To this end, we extend the idea of coordinate interleaved orthogonal design (CIOD) to our design. Specifically, in a DSM block, for a pair of modulated symbols drawn from a phase rotated constellation with angle θ, the real and imaginary parts of one modulated symbol is combined with the imaginary and real parts of the other modulated symbol. For example, the information matrix for $N_t = 2, 3$ with antenna activation orders (1 2) and (3 1 2) can be expressed by

$$\mathbf{X}_t = \begin{bmatrix} s_1^R + js_2^I & 0 \\ 0 & s_2^R + js_1^I \end{bmatrix}, \tag{3.53}$$

and

$$\mathbf{X}_t = \begin{bmatrix} 0 & s_2^R + js_3^I & 0 \\ 0 & 0 & s_3^R + js_1^I \\ s_1^R + js_2^I & 0 & 0 \end{bmatrix}, \tag{3.54}$$

respectively, where $\{s_i = s_i^R + js_i^I\}_{i=1}^4 \in \mathcal{X}^\theta$ and $\mathcal{X}^\theta$ denotes the rotated constellation $\mathcal{X}$ with angle θ.

Figure 3.10 shows the BER performances of the I-gray code order and gray code order with $N_t = 4$, $N_r = 2$, and $\theta = 15°$ in DSSK and DSM, respectively. We see that due to the improvement of diversity performance, the I-gray code order with 4-QAM outperforms the gray code order with either 4-QAM or BPSK in DSM at

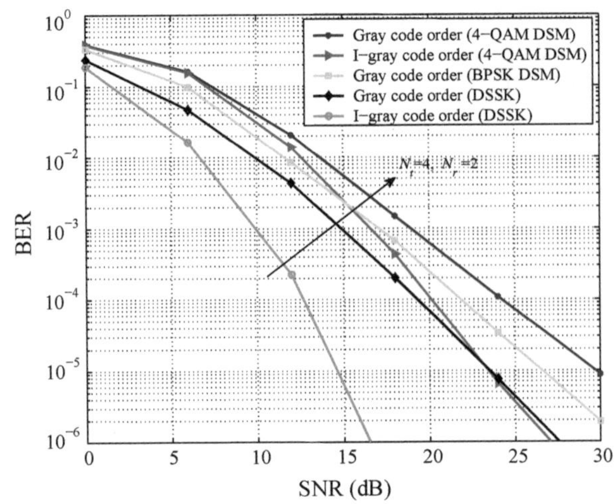

Fig. 3.10 BER performance between the gray code order and I-gray code order with $N_T = 4, N_R = 2$ in DSSK and DSM

high SNR. The I-gray code order with 4-QAM performs worse than the gray code order with BPSK at low SNR while better at high SNR. This is because the coding effect dominates the BER at low SNR while the diversity effect takes place at high SNR. In DSSK, I-gray code order still obtains an additional transmit diversity order at the price of one information bit loss for each transmission.

3.3 Precoding Aided Differential Spatial Modulation

In this section, we propose a novel MIMO scheme based on DSM termed as PDSM. By mapping information to the receiver-side space-time transmit blocks and conducting differential modulation, the PDSM scheme achieves reduced receiver-side complexity.

3.3.1 Transceiver Structure

We assume $N_t \geqslant N_r$ to ensure pre-coding design. In contrast to the traditional DSM, which conveys information via different transmitter-side space-time transmit blocks, the concept of PDSM is to activate specific receiver-side space-time receive blocks with the aid of pre-coding.

Design of PDSM Symbols

The first step of PDSM transmission is to map the information bits into space-time information blocks $\mathbf{X}_t \in \mathbb{C}^{N_r \times N_r}$, $(t = 1, 2 \ldots)$, which will be transmitted during time slot $N_r(t-1)+1$ to $N_r t$. The space-time information blocks are designed to satisfy the following two properties:

1. There is only one non-zero element in each column and row.
2. The non-zero elements are extracted from the M-PSK alphabet.

Thus, there are in all $N_r!$ different patterns of nonzero elements in $\mathbf{X}_t$, which can accommodate at most $\lfloor \log_2(N_r!) \rfloor$ space bits. We note that to utilize all possible $N_r!$ patterns to convey information, one can turn to the FBE scheme, which, however, may result in unacceptable encoding delay when the fraction is ratio of two large integers. Since every information block is transmitted over N_r time slots, the PDSM scheme achieves the spectrum efficiency measured by bps/Hz of

$$R_{PDSM} = \frac{1}{N_r} \left\lfloor \log_2(N_r!) \right\rfloor + \log_2(M). \tag{3.55}$$

Define the set of all possible $\mathbf{X}_t$ to be $\mathcal{G}$. To perform differential modulation, the DSM transmit blocks are constructed as follows. First, an arbitrary matrix $\mathbf{S}_0 \in \mathcal{G}$ is chosen to be the initial transmit block. Then, the t-th transmit block during time slot $N_r(t-1)+1$ to $N_r t$ is obtained according to

$$\mathbf{S}_t = \mathbf{S}_{t-1}\mathbf{X}_t, \quad t = 1, 2 \ldots. \tag{3.56}$$

Since the set $\mathcal{G}$ satisfies the closure property under multiplication:

$$\forall \mathbf{P}_1, \mathbf{P}_2 \in \mathcal{G}, \mathbf{P}_1\mathbf{P}_2 \in \mathcal{G}, \tag{3.57}$$

we have $\mathbf{S}_t \in \mathcal{G}$. We note that the constellation size M for the N_r non-zero entries in the information blocks are not necessarily to be the same. In this situation, if we define $\mathcal{G}$ to be the set where all the non-zero entries of its elements are extracted form the M_{max}-PSK constellation, where M_{max} refers to the maximum constellation size of the N_r nonzero symbols, the aforementioned property (3.57) still holds.

Pre-coding Design

For the t-th transmit block $\mathbf{S}_t$, the corresponding pre-coded block $\mathbf{O}_t \in \mathbb{C}^{N_t \times N_r}$ can be represented as

$$\mathbf{O}_t = \mathbf{PDS}_t, \tag{3.58}$$

where $\mathbf{P} \in \mathbb{C}^{N_t \times N_r}$ is the pre-coding matrix and $\mathbf{D} \in \mathbb{C}^{N_r \times N_r}$ is a diagonal power normalization matrix which guarantees that the transmitted power is averaged regardless of different transmit blocks. The i-th element of the main diagonal of $\mathbf{D}$ is equal to $d_i = \sqrt{1/\|\mathbf{p}_i\|^2}$, where $\mathbf{p}_i$ is the i-th column of the pre-coding matrix $\mathbf{P}$. $[\mathbf{O}_t]_{i,j}$ of the obtained pre-coded block represents the signal transmitted from the i-th transmit antenna at time slot $N_r(t-1)+j$.

If we assume perfect CSI at the transmitter, ZF pre-coding can be used to deliver desired signals to the receiver with perfect elimination of inter-channel interference. The ZF pre-coder can be formulated as:

$$\mathbf{P}_{ZF} = \mathbf{H}^H(\mathbf{H}\mathbf{H}^H)^{-1}. \tag{3.59}$$

Differential Detection

When ZF pre-coding is employed, the t-th *received block* at the receiver is equal to

$$\mathbf{Y}_t = \mathbf{D}\mathbf{S}_t + \mathbf{W}_t, \tag{3.60}$$

where $\mathbf{W}_t \in \mathbb{C}^{N_r \times N_r}$ is the independent identically distributed (i.i.d.) AWGN with $[\mathbf{W}_t]_{i,j}$
$\sim \mathcal{CN}(0, N_0)$. Given that the channel is quasi-static, by plugging (3.56) into (3.60), $\mathbf{Y}_t$ can be represented with $\mathbf{Y}_{t-1}$ as

$$\begin{aligned}\mathbf{Y}_t &= \mathbf{D}\mathbf{S}_{t-1}\mathbf{X}_t + \mathbf{W}_t \\ &= \mathbf{Y}_{t-1}\mathbf{X}_t + \mathbf{W}_t - \mathbf{W}_{t-1}\mathbf{X}_t.\end{aligned} \tag{3.61}$$

Accordingly, the non-coherent ML detector can be obtained as

$$\hat{\mathbf{X}}_t = \arg\max_{\forall \mathbf{X}_t \in \mathcal{G}} \mathrm{Tr}\left\{\Re\left\{\mathbf{Y}_t^H \mathbf{Y}_{t-1}\mathbf{X}_t\right\}\right\}. \tag{3.62}$$

3.3.2 ABEP Upper Bound Analysis

We employ the union bounding technique to analyze the upper bound on the ABEP for the case of $N_r = 2$.

When $N_r = 2$, only two information block patterns,

$$\begin{pmatrix} x_1 & 0 \\ 0 & x_2 \end{pmatrix}, \begin{pmatrix} 0 & x_2 \\ x_1 & 0 \end{pmatrix},$$

are available for transmission. Since $\Pr(\mathbf{X}_t = [\backslash]) = \Pr(\mathbf{X}_t = [/])$, an upper bound on the ABEP can be expressed as

$$P_e \leqslant \sum_{\mathbf{X}_t \in \mathcal{G}, \mathbf{X}_t = [\backslash]} \sum_{\widehat{\mathbf{X}}_t \in \mathcal{G}} \frac{N\left(\mathbf{X}_t \to \widehat{\mathbf{X}}_t\right) \Pr\left(\mathbf{X}_t \to \widehat{\mathbf{X}}_t\right)}{M^2\left(1 + 2\log_2(M)\right)}. \tag{3.63}$$

Lemma 2 *The instantaneous PEP of PDSM conditioned on* $\mathbf{D}^2$ *can be approximately calculated in terms of the Gaussian Q-function as*

$$Pr\left(\mathbf{X}_t \to \widehat{\mathbf{X}}_t | \mathbf{X}_t = [\backslash], \widehat{\mathbf{X}}_t = [\backslash], \mathbf{D}^2\right) = Q\left(\frac{\Re\{\alpha\} d_1^2 + \Re\{\beta\} d_2^2}{\sqrt{\left(|\alpha|^2 d_1^2 + |\beta|^2 d_2^2\right)\frac{1}{\gamma}}}\right) \tag{3.64}$$

and

$$Pr\left(\mathbf{X}_t \to \widehat{\mathbf{X}}_t | \mathbf{X}_t = [\backslash], \widehat{\mathbf{X}}_t = [/], \mathbf{D}^2\right) = Q\left(\sqrt{\frac{\left(d_1^2 + d_2^2\right)\gamma}{2}}\right), \tag{3.65}$$

respectively, where $\alpha = 1 - s_{2t-1}^* s_{2t-3} \hat{x}_{2t-1}$, $\beta = 1 - s_{2t}^* s_{2t-2} \hat{x}_{2t}$, *and* $\gamma = \frac{1}{N_0}$ *is the system SNR at each receive antenna;* s_t, x_t *and* $\hat{x}_t$ *are the transmit symbol, information symbol, and the detected symbol at time instant* t.

Proof The PEP can be written from (3.62) as

$$\Pr\left(\mathbf{X}_t \to \widehat{\mathbf{X}}_t\right) = \Pr\left(\mathrm{Tr}\left\{\Re\left\{\mathbf{Y}_t^H \mathbf{Y}_{t-1} \Delta\mathbf{X}_t\right\}\right\} > 0\right), \tag{3.66}$$

where $\Delta\mathbf{X}_t = \mathbf{X}_t - \widehat{\mathbf{X}}_t$.

If we assume that $\mathbf{S}_{t-1}$ is a diagonal matrix, when $\mathbf{X}_t$ and $\widehat{\mathbf{X}}_t$ are both diagonal matrices, $\mathrm{Tr}\left\{\mathbf{Y}_t^H \mathbf{Y}_{t-1} \Delta\mathbf{X}_t\right\}$ can be derived as

$$\begin{aligned}
\mathrm{Tr}\left\{\mathbf{Y}_t^H \mathbf{Y}_{t-1} \Delta\mathbf{X}_t\right\} = &\left(d_1 s_{2t-1}^* + n_{2t-1,1}^*\right)\left(d_1 s_{2t-3} + n_{2t-3,1}\right)\left(\hat{x}_{2t-1} - x_{2t-1}\right) \\
&+ n_{2t-1,2}^* n_{2t-3,2}\left(\hat{x}_{2t-1} - x_{2t-1}\right) \\
&+ \left(d_2 x_{2t}^* + n_{2t,2}^*\right)\left(d_2 x_{2t-2} + n_{2t-2,2}\right)\left(\hat{x}_{2t} - x_{2t}\right) \\
&+ n_{2t,1}^* n_{2t-2,1}\left(\hat{x}_{2t} - x_{2t}\right).
\end{aligned} \tag{3.67}$$

where $n_{t,i} \sim \mathcal{CN}(0, N_0)$, $i = 1, 2$, is the noise experienced at the i-th receive antenna at time instant t.

Recalling the relationship between information blocks and transmit blocks in (3.56), (3.67) can be further simplified as

$$\begin{aligned}
\mathrm{Tr}\left\{\mathbf{Y}_t^H \mathbf{Y}_{t-1} \Delta\mathbf{X}_t\right\} = &\alpha d_1^2 + \beta d_2^2 + \alpha d_1\left(s_{2t-3}^* n_{2t-3,1} + s_{2t-1} n_{2t-1,1}^*\right) \\
&+ \beta d_2\left(s_{2t-2}^* n_{2t-2,2} + s_{2t} n_{2t,2}^*\right) + o(n),
\end{aligned} \tag{3.68}$$

where $o(n)$ represents the higher order noise terms. If we ignore $o(n)$, $\mathrm{Tr}\left\{\mathbf{Y}_t^H \mathbf{Y}_{t-1} \Delta\mathbf{X}_t\right\}$ conditioned on specific $\mathbf{D}^2$ is a complex-valued Gaussian distributed RV with mean

$$\lambda_1 = \alpha d_1^2 + \beta d_2^2 \tag{3.69}$$

and variance

$$\sigma_1^2 = \frac{2}{\gamma}\left(\alpha^2 d_1^2 + \beta^2 d_2^2\right). \tag{3.70}$$

When $\mathbf{S}_{t-1}$ is an anti-diagonal matrix, $\mathrm{Tr}\left\{\mathbf{Y}_t^H \mathbf{Y}_{t-1} \Delta\mathbf{X}_t\right\}$ can be expressed in a similar form to (3.68) except for the switch between d_1 and d_2. If we notice that the probability of $\mathbf{S}_{t-1}$ being a diagonal matrix or an anti-diagonal matrix is equal to one half, and the statistics of d_1 and d_2 are the same (which will be clarified later), (3.64) can be obtained.

For the case of $\mathbf{X}_t$ being a diagonal matrix and $\widehat{\mathbf{X}}_t$ being an anti-diagonal matrix, if $\mathbf{S}_{t-1}$ is a diagonal matrix, we have

$$\begin{aligned}\mathrm{Tr}\left\{\mathbf{Y}_t^H\mathbf{Y}_{t-1}\Delta\mathbf{X}_t\right\} = & -d_1^2 - d_2^2 + d_1\left(s_{2t-1}^*\hat{x}_{2t-1}n_{2t-2,1} - s_{2t-1}^*x_{2t-1}n_{2t-3,1}\right.\\ & \left.+s_{2t-3}\hat{x}_{2t}n_{2t,1}^* - s_{2t-3}x_{2t-1}n_{2t-1,1}^*\right) + d_2\left(s_{2t}^*\hat{x}_{2t}n_{2t-3,2}\right.\\ & \left.-s_{2t}^*x_{2t}n_{2t-2,2} + s_{2t-2}\hat{x}_{2t-1}n_{2t-1,2}^* - s_{2t-2}x_{2t}n_{2t,2}^*\right) + o\left(n\right). \qquad (3.71)\end{aligned}$$

If we exclude higher order noise terms, $\mathrm{Tr}\left\{\mathbf{Y}_t^H\mathbf{Y}_{t-1}\Delta\mathbf{X}_t\right\}$ in (3.71) conditioned on specific $\mathbf{D}^2$ is also a complex-valued Gaussian RV with mean

$$\lambda_2 = -d_1^2 - d_2^2 \qquad (3.72)$$

and variance

$$\sigma_2^2 = \frac{4}{\gamma}\left(d_1^2 + d_2^2\right). \qquad (3.73)$$

Therefore, (3.65) can be obtained for the same reason given before. □

As can be seen from (3.64) and (3.65), the upper bound depends on RVs $d_i^2, i = 1, 2$. By the definition of $\mathbf{D}$, we have

$$d_i^2 = \frac{1}{\left[\left(\mathbf{HH}^H\right)^{-1}\right]_{i,i}}, \qquad (3.74)$$

where $\mathbf{HH}^H$ is the wishart random matrix, and $d_i^2, i = 1, 2$, have the PDF of

$$f_{d_i^2}\left(x\right) = \frac{1}{\Gamma\left(L\right)}x^{L-1}e^{-x},\ x > 0. \qquad (3.75)$$

Now the ABEP upper bound can be calculated from (3.63). For $M > 2$, the statistics of the RVs inside the Gaussian Q-functions in (3.64) and (3.65) are too complicated to characterize. However, a closed-form expression for the upper bound on the ABEP is available for BPSK signaling.

The error events for BPSK modulation can be explicitly defined as

- $\bullet\mathrm{E1}: \mathbf{X}_t = [\backslash], \widehat{\mathbf{X}}_t = [\backslash], \hat{x}_{2t-1} \neq x_{2t-1}, \hat{x}_{2t} = x_{2t};$
- $\bullet\mathrm{E2}: \mathbf{X}_t = [\backslash], \widehat{\mathbf{X}}_t = [\backslash], \hat{x}_{2t-1} = x_{2t-1}, \hat{x}_{2t} \neq x_{2t};$
- $\bullet\mathrm{E3}: \mathbf{X}_t = [\backslash], \widehat{\mathbf{X}}_t = [\backslash], \hat{x}_{2t-1} \neq x_{2t-1}, \hat{x}_{2t} \neq x_{2t};$
- $\bullet\mathrm{E4}: \mathbf{X}_t = [\backslash], \widehat{\mathbf{X}}_t = [/], \hat{x}_{2t-1} = x_{2t-1}, \hat{x}_{2t} = x_{2t};$
- $\bullet\mathrm{E5}: \mathbf{X}_t = [\backslash], \widehat{\mathbf{X}}_t = [/], \hat{x}_{2t-1} \neq x_{2t-1}, \hat{x}_{2t} = x_{2t};$
- $\bullet\mathrm{E6}: \mathbf{X}_t = [\backslash], \widehat{\mathbf{X}}_t = [/], \hat{x}_{2t-1} = x_{2t-1}, \hat{x}_{2t} \neq x_{2t};$
- $\bullet\mathrm{E7}: \mathbf{X}_t = [\backslash], \widehat{\mathbf{X}}_t = [/], \hat{x}_{2t-1} \neq x_{2t-1}, \hat{x}_{2t} \neq x_{2t}.$

Since the RVs of d_i^2, $i = 1, 2$, have the same PDF and the BPSK symbols are drawn form $\{-1, +1\}$, in light of (3.64) and (3.65), we can group the error events into 3 different types and write their PEPs as

$$\Pr\left(\mathbf{X}_t \to \widehat{\mathbf{X}}_t | \mathrm{E1}, \mathbf{D}^2\right) = \Pr\left(\mathbf{X}_t \to \widehat{\mathbf{X}}_t | \mathrm{E2}, \mathbf{D}^2\right) = Q\left(\sqrt{d_1^2 \gamma}\right), \tag{3.76}$$

$$\Pr\left(\mathbf{X}_t \to \widehat{\mathbf{X}}_t | \mathrm{E3}, \mathbf{D}^2\right) = Q\left(\sqrt{\left(d_1^2 + d_2^2\right)\gamma}\right) \tag{3.77}$$

and

$$\begin{aligned}\Pr\left(\mathbf{X}_t \to \widehat{\mathbf{X}}_t | \mathrm{E4}, \mathbf{D}^2\right) &= \Pr\left(\mathbf{X}_t \to \widehat{\mathbf{X}}_t | \mathrm{E5}, \mathbf{D}^2\right) = \Pr\left(\mathbf{X}_t \to \widehat{\mathbf{X}}_t | \mathrm{E6}, \mathbf{D}^2\right) \\ &= \Pr\left(\mathbf{X}_t \to \widehat{\mathbf{X}}_t | \mathrm{E7}, \mathbf{D}^2\right) = Q\left(\sqrt{\frac{\left(d_1^2 + d_2^2\right)\gamma}{2}}\right),\end{aligned} \tag{3.78}$$

respectively.

By averaging the Q-function over all realizations of d_1^2 using (3.75), we can bound the PEP of (3.76) as

$$\begin{aligned}\Pr\left(\mathbf{X}_t \to \widehat{\mathbf{X}}_t | \mathrm{E1}\right) &= \Pr\left(\mathbf{X}_t \to \widehat{\mathbf{X}}_t | \mathrm{E2}\right) \\ &= E_{d_1^2}\left\{Q\left(\sqrt{d_1^2 \gamma}\right)\right\} \leqslant \frac{1}{2}\left(\frac{\gamma}{2} + 1\right)^{-L},\end{aligned} \tag{3.79}$$

where the Chernoff bound $Q(x) \leqslant \frac{1}{2} e^{-x^2/2}$ has been used.

For (3.77) and (3.78), the evaluation requires the joint PDF of the correlated RVs d_1^2 and d_2^2. Given that d_1^2 and d_2^2 are correlated and distributed as Gamma$(L, 1)$, their joint PDF according to [[11], Eq. (8.3)] is the Kibble's bivariate gamma distribution

$$f_{d_1^2, d_2^2}(x, y) = \frac{(1-\rho)^{-L}}{\Gamma(L)} \sum_{k=0}^{\infty} a_k (xy)^{L+k-1} e^{-\frac{x+y}{1-\rho}}, \tag{3.80}$$

where $a_k = \rho^k / \left[\Gamma(L+k)(1-\rho)^{2k} k!\right]$, and $\rho = E\{(d_1^2 - L)(d_2^2) - L\}/L$ is the Pearson product-moment correlation coefficient that can be estimated for a given system configuration.

Following the same procedure in [[12], Eq. (63)], by averaging (3.77) with (3.80), we have

$$\begin{aligned}\Pr\left(\mathbf{X}_t \to \widehat{\mathbf{X}}_t | \mathrm{E3}\right) &= E_{\mathbf{D}^2}\left\{Q\left(\sqrt{\left(d_1^2 + d_2^2\right)\gamma}\right)\right\} \\ &\leqslant \frac{(1-\rho)^{-L}}{2 t_1^{2L}} \, {}_1F_0\left(L; ; \frac{\rho}{(1-\rho)^2 t_1^2}\right),\end{aligned} \tag{3.81}$$

where $t_1 = \gamma/2 + 1/(1-\rho)$ and

$$ {}_1F_0\left(L;;x\right) = \sum_{k=0}^{+\infty} \frac{(L)^{(k)}}{k!} x^k \tag{3.82} $$

is the generalized hypergeometric function as defined in [[5], p. 1010] with $(L)^{(k)} = L\left(L+1\right)\cdots\left(L+k+1\right)$ denoting the Pochhammer symbol.

Similarly, the PEP in (3.78) can be averaged by (3.80) and written as:

$$ \begin{aligned} \Pr\left(\mathbf{X}_t \rightarrow \widehat{\mathbf{X}}_t|\text{E4}\right) &= \Pr\left(\mathbf{X}_t \rightarrow \widehat{\mathbf{X}}_t|\text{E5}\right) = \Pr\left(\mathbf{X}_t \rightarrow \widehat{\mathbf{X}}_t|\text{E6}\right) \\ &= \Pr\left(\mathbf{X}_t \rightarrow \widehat{\mathbf{X}}_t|\text{E7}\right) \\ &= \mathrm{E}_{\mathbf{D}^2}\left\{Q\left(\sqrt{\frac{\left(d_1^2+d_2^2\right)\gamma}{2}}\right)\right\} \\ &\leqslant \frac{(1-\rho)^{-L}}{2t_2^{2L}}\,{}_1F_0\left(L;;\frac{\rho}{(1-\rho)^2 t_2^2}\right), \end{aligned} \tag{3.83} $$

where $t_2 = \gamma/4 + 1/(1-\rho)$.

Finally, by plugging (3.79), (3.81) and (3.83) into (3.63), we have the closed-form ABEP upper bound for BPSK signaling as

$$ \begin{aligned} P_e \leqslant & \frac{1}{3}\left[\left(\frac{\gamma}{2}+1\right)^{-L} + \frac{(1-\rho)^{-L}}{t_1^{2L}}\,{}_1F_0\left(L;;\frac{\rho}{(1-\rho)^2 t_1^2}\right)\right. \\ & \left. + \frac{4(1-\rho)^{-L}}{t_2^{2L}}\,{}_1F_0\left(L;;\frac{\rho}{(1-\rho)^2 t_2^2}\right)\right]. \end{aligned} \tag{3.84} $$

3.3.3 Performance Evaluation

Monte Carlo simulations are conducted under quasi-static Rayleigh fading channels to examine the performance of PDSM and verify the theoretical analysis. The SNR is defined as $\gamma = P_s/N_0 = 1/N_0$, where the signal power P_s is set to be 1 for the use of the normalization matrix $\mathbf{D}$ and the normalization of M-PSK constellation points.

Simulation Versus Theoretical Results

Before comparison with the benchmark, Fig. 3.11 demonstrates the simulation results and theoretical analysis of our proposed PDSM architecture with 2 receive antennas and BPSK. We note that due to the ignorance of higher order noises and the bounding techniques, our theoretical analysis of (3.84) is relatively loose. However, if we add a 1.7 dB offset to the theoretical results, the simulation curve of the considered system

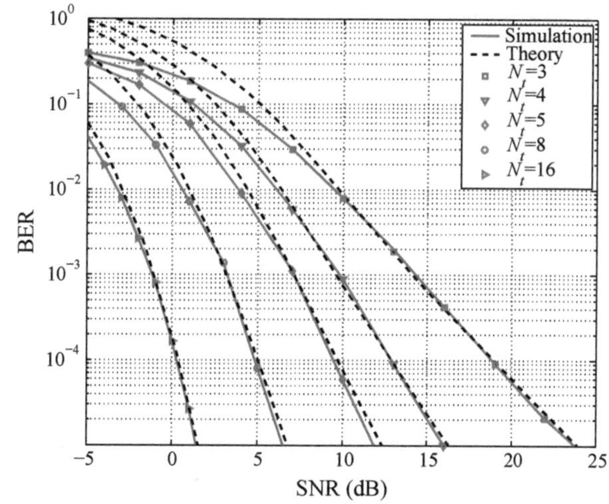

Fig. 3.11 BER performance of PDSM with 2 receive antennas and BPSK

setups perfectly matches the theoretical curve. As can be seen from the figure, the slope of the curve increases with the number of transmit antennas.

PDSM Versus PSM

Figure 3.12 compares the BER performance of PDSM and PSM under identical system configuration using $N_r = 2, 4$ receive antennas. For $N_r = 2$, The PSM system applies BPSK constellation, while the PDSM system uses mixed BPSK and QPSK constellation to reach the same spectrum efficiency of 2 bps/Hz. For $N_r = 4$, The PSM system applies BPSK constellation, while the PDSM system uses QPSK to reach the same spectrum efficiency of 3 bps/Hz. One can see from the figure that although higher modulation order is used in PDSM to match the data rate, the BER performance loss induced by the differential scheme in comparison with PSM is no more than 3 dB. Moreover, the performance losses between PDSM and PSM both reduce from about 3 to 2 dB with the transmit antenna number increasing from $N_t = 4$ to $N_t = 16$ for either $N_r = 2$ or $N_r = 4$.

PDSM with Multiple Receive Antennas

Figure 3.13 compares the BER performance of PDSM with $N_r = 3$ and $N_r = 4$ at the fixed data rate of 3 bps/Hz. 4-receive-antenna system uses QPSK in all 4 nonzero elements of each information block and reaches a space bit data rate of 1 bps/Hz, while 3-receive-antenna system uses two QPSK and one 8-PSK constellation to match the data rate and achieves a space bit data rate of 2/3 bps/Hz. As can be seen from the simulation results, the 4-receive-antenna system outperforms 3-receive-antenna system when equipped with more than six transmit antennas at a relatively high SNR region. It can be explained with the higher constellation order applied in 3-receive-antenna system. However, 3-receive-antenna system exhibits better performance for $N_t < 6$, and in the low SNR region for all considered transmit antenna configurations. It is because the space bits are more vulnerable in face of poor channel state or less

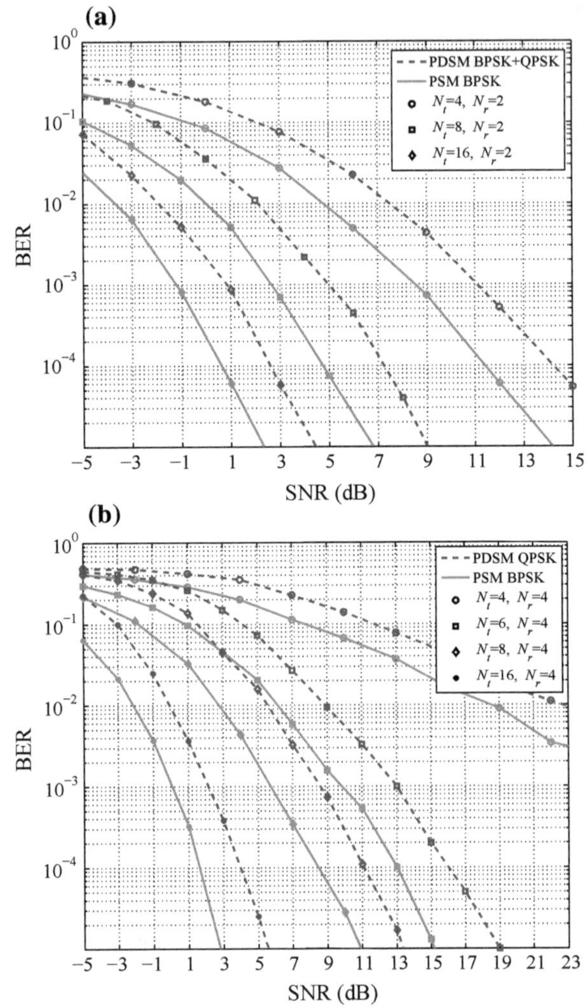

Fig. 3.12 Comparison of BER performance between PDSM and PSM with: **a** 2 receive antennas at 2 bps/Hz and **b** 4 receive antennas at 3 bps/Hz

robust pre-coding matrices. When space bit error occurs, the constellation signal bits in the wrong space-time block pattern will be determined randomly by the noise.

This observation reveals a trade-off between space bits and constellation bits. We can make proper optimization at the receiver according to the transmit antenna configuration and system SNR, e.g., when fewer transmit antennas are equipped, the receiver may deactivate certain antenna and correspondingly increase the constellation order to achieve the optimal performance without sacrificing data rate.

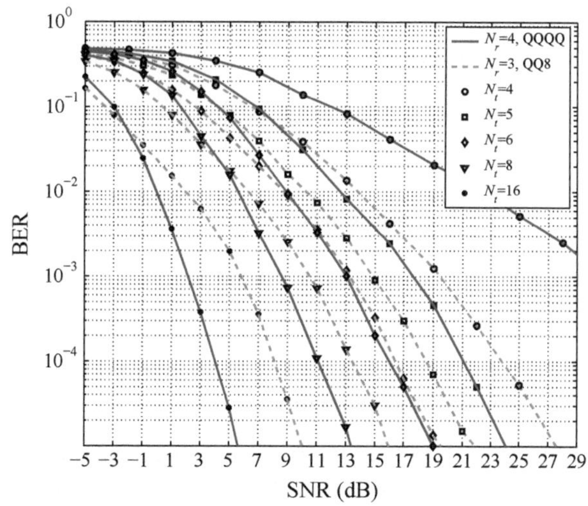

Fig. 3.13 Comparison of BER performance of PDSM with $N_r = 3$ and $N_r = 4$ at 3 bps/Hz

3.4 Applications to Cooperative Communications

In this section, we propose a dual-hop hybrid DSM (DH-HDSM) architecture, where a PDSM architecture in the first hop and a DSM architecture in the second hop are formed.

3.4.1 System Model

The system model of DH-HDSM is depicted in Fig. 3.14. We consider a wireless communication system with a source node (SN) of N_S antennas and a destination node (DN) of N_D antennas. The direct link between the SN and the DN is assumed to be too weak to support high quality transmission and thus, N_R single antenna relay nodes (RNs) serve to deliver the message from the SN to the DN. The RNs are considered to be either connected with an error-free backhaul link or in a distributed form, which will be illustrated later. To enable linear pre-coding, $N_S \geq N_R$ is assumed and the channel is supposed to be quasi-static during the transmission. Also, the SN is assumed to have perfect CSI of the first hop.

Design of the Information and Transmit Blocks

The speciality of DH-HDSM lies in the design of the information blocks $\mathbf{X}_t \in \mathbb{C}^{N_R \times N_R}$, $t = 1, 2, 3 \ldots$, which will be transmitted from the SN to the RNs during time slots $(2t - 2)N_R + 1 \sim (2t - 1)N_R$. To ensure differential modulation, each information block contains N_R non-zero elements and satisfies the properties of:

1. There is only one non-zero element in each column and row.

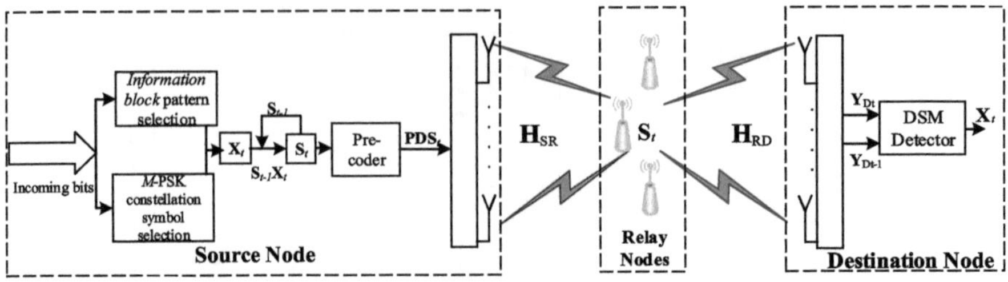

Fig. 3.14 System model of DH-HDSM

2. The non-zero elements are extracted from the normalized M-PSK alphabet, where M denotes the cardinality.

In consideration of property 1), we notice that there are in total $N_R!$ different patterns of the non-zero elements. Thus, $k_1 = \lfloor \log_2(N_R!) \rfloor$ information bits, which will be referred to as space bits, can be chosen to select a specific pattern of the information blocks. Then, by mapping another $k_2 = N_R \log_2(M)$ constellation bits to the M-PSK constellation to form the N_R non-zero elements, we obtain the corresponding $\mathbf{X}_t \in \mathcal{G}$, with $\mathcal{G}$ defined to be the set of all possible $\mathbf{X}_t$. An arbitrary element can be chosen from $\mathcal{G}$ as the initial transmit block $\mathbf{S}_0$. Then, the subsequent transmit blocks can be obtained according to $\mathbf{S}_t = \mathbf{S}_{t-1}\mathbf{X}_t$. Since the pattern of the non-zero elements inside $\mathbf{X}_t$ can be regarded as the permutation of the columns of the identity matrix, and the non-zero elements are drawn from PSK constellations, it is evident that the information blocks satisfy the closure property under multiplication,

$$\forall \mathbf{P}_1, \mathbf{P}_2 \in \mathcal{G}, \mathbf{P}_1\mathbf{P}_2 \in \mathcal{G}. \tag{3.85}$$

Thus, the *transmit blocks* are also drawn from $\mathcal{G}$.

Transmission of the First Hop

The received signal at the RNs can be expressed as

$$\mathbf{Y}_{R_t} = \mathbf{H}_{SR}\mathbf{PDS}_t + \mathbf{W}_{R_t}, \tag{3.86}$$

where $\mathbf{Y}_{R_t} \in \mathbb{C}^{N_R \times N_R}$ collects the received signals at the RNs during time slots $(2t-2)N_R + 1 \sim (2t-1)N_R$, and $\mathbf{H}_{SR} \in \mathbb{C}^{N_R \times N_T}$ represents the channel matrix between the SN and the RNs, whose elements are assumed to be i.i.d. RVs $\sim \mathcal{CN}(0, 1)$. $\mathbf{P} \in \mathbb{C}^{N_T \times N_R}$ and $\mathbf{D} \in \mathbb{C}^{N_R \times N_R}$ are the pre-coding matrix and the power normalization matrix, respectively. The i-th diagonal element of $\mathbf{D}$ is equal to $d_i = \sqrt{1/\|\mathbf{p}_i\|^2}$, where $\mathbf{p}_i$ is the i-th column of the pre-coding matrix $\mathbf{P}$ with $i \in \{1, \ldots, N_R\}$. By recalling the properties of the transmit blocks, we can easily find that the power normalization matrix ensures unity transmit power ($P_s = 1$) at each time instant. We note that a simple power normalization factor β can also be used here as $\beta = \sqrt{N_R/\mathrm{Tr}\{\mathbf{PP}^H\}}$. However, it simply sets a limit to the average transmit power over all possible transmit symbols, such that the instantaneous transmit power may still be possible to reach a relatively high level, which requires the amplifiers to have much wider dynamic range [13]. Thus, in consideration of practical deployment, we choose the power normalization matrix to maintain constant transmit power and relax the hardware restrictions. Finally, the last term $\mathbf{W}_{R_t} \in \mathbb{C}^{N_R \times N_R}$ in (3.86) represents the i.i.d. AWGNs experienced at the RNs with its elements following $\mathcal{CN}(0, \sigma_R^2)$. The SNR of the first hop is defined as $\gamma_{SR} = P_s/\sigma_R^2 = 1/\sigma_R^2$.

We consider ZF pre-coding. The ZF pre-coding matrix is the pseudo-inverse of the channel matrix $\mathbf{H}_{SR}$, which is equal to

$$\mathbf{P}_{ZF} = \mathbf{H}_{SR}^H \left(\mathbf{H}_{SR}\mathbf{H}_{SR}^H\right)^{-1}. \tag{3.87}$$

Under the assumption of ZF pre-coding, the t-th received block at the RNs can be given by

$$\mathbf{Y}_{R_t} = \mathbf{D}\mathbf{S}_t + \mathbf{W}_{R_t}, \tag{3.88}$$

with its (i, j)-th element representing the received signal at the i-th RN at time slot $(2t-2)N_R+j$. In this way, the PDSM architecture is formed for the first hop. We note that due to pre-coding, the transmit antennas at the SN are active simultaneously. However, at the RNs, since $\mathbf{D}$ is a diagonal matrix, the pattern of the non-zero elements in the collection of the received signals at the RNs is exactly in the form of DSM transmit blocks. Consequently, if we ignore the influence of noise, only one RN receives signal at each time slot and each RN receives signal once during the transmission. In other words, the space bits are mapped to the received signal patterns at the RNs.

Centralized Detection Algorithm

If the RNs are connected with a backhaul link, ML centralized detection (CD) algorithm can be employed by jointly using the information from all RNs. Since the channel is assumed to be quasi-static, recalling the relationship between transmit blocks and information blocks, we notice that the t-th received block in (3.88) can be represented by the $(t-1)$-th received block as

$$\mathbf{Y}_{R_t} = \mathbf{Y}_{R_{t-1}}\mathbf{X}_t - \mathbf{W}_{R_{t-1}}\mathbf{X}_t + \mathbf{W}_{R_t}. \tag{3.89}$$

Thus, we are able to apply differential detection to retrieve the t-th information block according to

$$\begin{aligned}\tilde{\mathbf{X}}_t &= \arg\min_{\forall \mathbf{X}_t \in \mathcal{G}} \left\{\left\|\mathbf{Y}_{R_t} - \mathbf{Y}_{R_{t-1}}\mathbf{X}_t\right\|^2\right\} \\ &= \arg\max_{\forall \mathbf{X}_t \in \mathcal{G}} \operatorname{Tr}\left\{\Re\left\{\mathbf{Y}_{R_t}^H \mathbf{Y}_{R_{t-1}}\mathbf{X}_t\right\}\right\}.\end{aligned} \tag{3.90}$$

As can be seen from the above equation, the ML CD algorithm only requires the knowledge of the present and previous received blocks and the power normalization matrix $\mathbf{D}$ is no longer needed.

Given that the initial transmit block $\mathbf{S}_0$ is known prior at the RNs, the corresponding transmit block $\widetilde{\mathbf{S}}_t$ to be relayed to the DN can be calculated by

$$\widetilde{\mathbf{S}}_t = \widetilde{\mathbf{S}}_{t-1}\widetilde{\mathbf{X}}_t, \tag{3.91}$$

with its (i, j)-th entry denoting the signal to be transmitted from the i-th relay at time slot $(2t-1)N_R + j$ during the second hop. Since $\widetilde{\mathbf{X}}_t$ and $\widetilde{\mathbf{S}}_t$ are both drawn from $\mathcal{G}$, each RN is activated once, and only one RN is active at any time slot during the second hop.

Distributed Detection Algorithms

In some cases, the backhaul link between RNs comes at a very high price, because usually, the individual RNs in the virtual MIMO (VMIMO) scenarios are battery-powered mobile terminals, the backhual links will cost the wireless spectrum and battery resources, which are crucial to the communication quality and service time of the terminals. Thus, the removal of backhaul links and the resultant distributed detection (DD) protocol are much more practical. On the one hand, with the DD protocol, the RNs can be released from the heavy battery resource demand and extend the service duration. On the other hand, the saved spectrum can also be utilized to boost the system capacity. Since the RNs detect their own signals without assistance of each other, to illustrate the DD algorithms, we first write down the received signal vectors at the i-th RN from (3.88) as

$$\mathbf{y}_{R_t,i} = d_i\mathbf{s}_{t,i} + \mathbf{w}_{R_t,i}, \tag{3.92}$$

where $\mathbf{y}_{R_t,i}$, $\mathbf{s}_{t,i}$ and $\mathbf{w}_{R_t,i}$ represent the i-th row of $\mathbf{Y}_{R_t}$, $\mathbf{S}_t$ and $\mathbf{W}_{R_t}$, respectively. In consideration of the property that there exists only one non-zero elements in every row of $\mathbf{S}_t$, we can rewrite $\mathbf{s}_{t,i} \in \mathcal{X}$ as $\mathbf{s}_{t,i} = \mathbf{e}_i c_i$, where $\mathbf{e}_i = [0, \ldots, 1, \ldots, 0]$ is a $1 \times N_R$ vector drawn from the i-th row of the identity matrix $\mathbf{I}_{N_R}$ with $i \in \{1, \ldots, N_R\}$, $c_i \in \{s_1, \ldots, s_M\}$ represents the symbol selected from the M-PSK signal constellation, and $\mathcal{X}$ is the collection of all possible $\mathbf{s}_{t,i}$. The simulation results will show that by properly using the time domain properties of the received signal vector, the proposed DD detectors could harvest the performance gain provided by the time-domain joint detection.

In the following, we will introduce the optimal ML detector and the simplified near-optimal one for the RNs respectively.

(a) Optimal ML Detector

From (3.92), the optimal ML detection at the j-th RN can be carried out according to

$$\widetilde{\mathbf{s}}_{t,i} = \arg\min_{\forall \mathbf{s}_{t,i} \in \mathcal{X}} \left\| \mathbf{y}_{R_t,i} - d_i\mathbf{s}_{t,i} \right\|^2. \tag{3.93}$$

We note that the above equation can be further simplified due to the fact that d_i is a real value, which means that the pre-coding method neutralizes the phase shift caused by the channels and the received signal is simply a copy of the original symbol with an scaling factor of d_i. Using the fact that c_i is an M-PSK symbol, the above equation can be represented in a simpler form of

$$\widetilde{\mathbf{s}}_{t,i} = \arg\max_{\forall \mathbf{s}_{t,i} \in \mathcal{X}} \Re\left\{\mathbf{y}^*_{R_t,i}\mathbf{s}_{t,i}\right\}. \tag{3.94}$$

Thus, incoherent ML detection can be done at the RNs without the knowledge of d_i. Since only one non-zero elements exists in $\mathbf{s}_{t,i}$, it can be easily obtained that the total search complexity of the above ML detector is as low as $\mathcal{O}(MN_R)$ per RN.

(b) Low-Complexity Near-Optimal Detector

To further reduce the computational expense at the RNs, we propose a two-stage near-optimal DD algorithm, which takes the time interval with the largest received power as the active one, and then conducts ML detection to obtain the the most likely M-PSK symbol. The process of the proposed detector at the i-th RN can be demonstrated as

$$\begin{aligned} \widetilde{j}_i &= \arg\max_{j_i=\{1,\ldots,N_R\}} \left|\mathbf{y}_{R_t,i}\mathbf{e}^T_{j_i}\right|, \\ \widetilde{c}_i &= \arg\max_{\forall c_i \in \{s_1,\ldots,s_M\}} \Re\left\{\mathbf{y}_{R_t,i}\mathbf{e}^T_{\widetilde{j}_i} c^*_i\right\}. \end{aligned} \tag{3.95}$$

It is clear from (3.95) that the search complexity of the sub-optimal detector is reduced to M per RN, which is $1/N_R$ of that of the ML detector.

Obviously, both DD algorithms can be done at each RN independently and ensure that each RN detects only one active time slot. By concatenating the detected vectors obtained by the DD algorithms, the t-th detected transmit block can be written as $\widetilde{\mathbf{S}}_t = \left[\widetilde{c}_1\mathbf{e}^T_{\widetilde{j}_1}, \cdots, \widetilde{c}_{N_R}\mathbf{e}^T_{\widetilde{j}_{N_R}}\right]^T$, which will be forwarded to the DN with its elements defined exactly the same as in (3.91). We note that because the RNs determine the received symbols in a distributed form, corresponding vectors in the received transmit blocks can be directly obtained by the detector. This is quite different from the CD, which extracts the information block $\widetilde{\mathbf{X}}_t$ via differential demodulation and calculates the corresponding transmit block $\widetilde{\mathbf{S}}_t$ afterwards according to the differential modulation criterion.

Transmission and Detection of the Second Hop

After the detection in an either centralized or distributed form, the RNs obtain the transmit block $\widetilde{\mathbf{S}}_t$ to be relayed to the DN. Then, during the time slots $(2t-1)N_R+1 \sim 2tN_R$, the RNs forward $\widetilde{\mathbf{S}}_t$ to the DN, resulting in the received block

$$\mathbf{Y}_{D_t} = \mathbf{H}_{RD}\widetilde{\mathbf{S}}_t + \mathbf{W}_{D_t}, \tag{3.96}$$

where $\mathbf{Y}_{D_t} \in \mathbb{C}^{N_D \times N_R}$ collects the received signals at the DN, $\mathbf{H}_{RD} \in \mathbb{C}^{N_D \times N_R}$ represents the channel matrix between the RNs and the DN, whose elements are assumed to be i.i.d. RVs $\sim \mathcal{CN}(0, 1)$, and $\mathbf{W}_{D_t} \in \mathbb{C}^{N_D \times N_R}$ denotes the i.i.d. AWGNs experienced at the DN with its elements following $\mathcal{CN}(0, \sigma_D^2)$. Similar to that of the first hop, the SNR of the second hop is defined as $\gamma_{RD} = P_r/\sigma_D^2 = 1/\sigma_D^2$, where P_r is the average transmit power at the RNs and is equal to 1 because the proposed CD and DD algorithms ensure that each RN is activated precisely once during each transmit block. Since the transmission of each block requires $2N_R$ time instants, we have the data rate of DH-HDSM measured in bps/Hz as $k = (k_1 + k_2)\big/2N_R$.

Finally, the optimal differential ML detection can be applied at the DN according to

$$\widehat{\mathbf{X}}_t = \underset{\forall \mathbf{X}_t \in \mathcal{G}}{\arg\min} \left\| \mathbf{Y}_{D_t} - \mathbf{Y}_{D_{t-1}} \mathbf{X}_t \right\|^2. \tag{3.97}$$

We note that for CD at the RNs, $\widetilde{\mathbf{S}}_t$ is a legal transmit block. The transmission of the second hop is therefore identical to that of a DSM system. However, as mentioned before, although both proposed DD algorithms ensure that each RN detects one non-zero elements during the first hop transmission, the relayed block may be illegal because it's possible that at a single time slot, multiple RNs are activated, or all RNs remain silent. Since the DN is unaware of the error detection at the RNs, the detector may suffer from a performance penalty caused by the illegal transmit blocks. Error-aware detection at the DN is believed to be able to mitigate the performance loss, which, however, is at the cost of other system expenses. In order to achieve low complexity, we decide to discard error correction at the DN and assume $\widetilde{\mathbf{S}}_t$ to be always legal. As will be illustrated in the simulations, the performance loss caused by the illegal transmit blocks is negligible at high SNR.

3.4.2 ABEP Upper Bound Analysis

The well-known union bounding technique is employed to evaluate the ABEP of the proposed scheme. Our performance analysis focuses on ZF pre-coding.

Average Bit Error Probability for CD with Two RNs and BPSK

When CD is employed at the RNs, the two hops of DH-HDSM can be treated as independent PDSM and DSM transmissions. Thus, the end-to-end ABEP can be expressed in a similar form as in [4], which is given by

$$P_{SD}\left(\gamma_{SR}, \gamma_{RD}\right) = P_R\left(\gamma_{SR}\right) + P_D\left(\gamma_{RD}\right) - P_R\left(\gamma_{SR}\right) P_D\left(\gamma_{RD}\right), \tag{3.98}$$

where $P_R\left(\gamma_{SR}\right)$ and $P_D\left(\gamma_{RD}\right)$ represent the ABEPs at the RNs (PDSM) and the DN (DSM), respectively.

To arrive at a closed-form expression of the ABEP, we focus on the situation where two RNs and BPSK are involved. We begin with the derivation of the APEP of the first hop (PDSM).

Lemma 3 *The ABEP upper bound of the first hop with 2 RNs, BPSK and CD is given by*

$$
\begin{aligned}
P_R(\gamma_{SR}) \leq & \frac{1}{3}\left[2(1-f(\gamma_{SR}))^L \sum_{l=0}^{L-1} \frac{1}{2^{L+l}} \mathrm{C}(L-1+l, l) \right. \\
& \times (1+f(\gamma_{SR}))^l + \frac{(1-\rho)^{-L}}{t_1^{2L}} {}_1F_0\left(L;;\frac{\rho}{(1-\rho)^2 t_1^2}\right) \\
& \left. + \frac{4(1-\rho)^{-L}}{t_2^{2L}} {}_1F_0\left(L;;\frac{\rho}{(1-\rho)^2 t_2^2}\right)\right], \qquad (3.99)
\end{aligned}
$$

where $f(\gamma_{RD}) = \sqrt{\gamma_{RD}/(2+\gamma_{RD})}$, $L = N_S - N_R + 1$, $t_1 = \gamma_{SR}/2 + 1/(1-\rho)$, *and* $t_2 = \gamma_{SR}/4 + 1/(1-\rho)$ *with* $\rho = E\{(d_1^2 - L)(d_2^2) - L\}/L$ *being the Pearson product-moment correlation coefficient.*

Proof This proof is similar to that of Lemma 2, and thus is omitted here for brevity. □

The evaluation of the second hop (DSM) has been well discussed in Sect. 3.3.2, which can be applied in our analysis as

Lemma 4 *The ABEP upper bound of the second hop with 2 RNs, BPSK and CD is given by*

$$
\begin{aligned}
P_D(\gamma_{RD}) \leqslant & (1-f(\gamma_{RD}))^{N_D} \sum_{l=0}^{N_D-1} \frac{1}{2^{N_D-1+l}} \mathrm{C}(N_D-1+l, l)(1+f(\gamma_{RD}))^l \\
& + \sum_{l=0}^{2N_D-1} \frac{\mathrm{C}(2N_D-1+l, l)}{3\cdot 2^{2N_D-1+l}} \left[(1-f(\gamma_{RD}))^{2N_D}(1+f(\gamma_{RD}))^l \right. \\
& \left. + 2(1-f(\gamma_{RD}/2))^{2N_D}(1+f(\gamma_{RD}/2))^l\right]. \qquad (3.100)
\end{aligned}
$$

Now, with (3.99) and (3.100), we can complete the derivation of (3.98).

Average Bit Error Probability for DD

We provide the ABEP analysis of the proposed scheme when the RNs apply the DD algorithms for arbitrary N_S, N_R and N_D and arbitrary constellation size. Since the transmitted block $\widetilde{\mathbf{S}}_t$ might be an illegal transmit block, (3.98) does not hold any more. Thus, we evaluate the end-to-end ABEP via

$$
P_{SD}(\gamma_{SR}, \gamma_{RD}) \leqslant \sum_{\mathbf{X}_t\in\mathcal{G}, \mathbf{X}_t=[\backslash]} \sum_{\widehat{\mathbf{X}}_t\in\mathcal{G}} \frac{N\left(\mathbf{X}_t \to \widehat{\mathbf{X}}_t\right) P_{SD}\left(\mathbf{X}_t \to \widehat{\mathbf{X}}_t, \gamma_{SR}, \gamma_{RD}\right)}{M^2\left(1+2\log_2(M)\right)}, \qquad (3.101)
$$

where $P_{SD}\left(\mathbf{X}_t \to \widehat{\mathbf{X}}_t, \gamma_{SR}, \gamma_{RD}\right)$ is the end-to-end PEP when $\mathbf{X}_t$ is transmitted at the SN and $\widehat{\mathbf{X}}_t$ is erroneously detected at the DN, which can be represented in another form by taking $\widetilde{\mathbf{X}}_t$ into account:

$$P_{SD}\left(\mathbf{X}_t \to \widehat{\mathbf{X}}_t, \gamma_{SR}, \gamma_{RD}\right) = \sum_{\widetilde{\mathbf{X}}_t \in \mathfrak{A}} P_{SR}\left(\mathbf{X}_t \to \widetilde{\mathbf{X}}_t, \gamma_{SR}\right) P_{RD}\left(\widetilde{\mathbf{X}}_t \to \widehat{\mathbf{X}}_t, \gamma_{RD}\right), \quad (3.102)$$

where $\widetilde{\mathbf{X}}_t \in \mathfrak{A}$ includes all the possible detected legal and illegal information blocks at the RNs, $P_{SR}\left(\mathbf{X}_t \to \widetilde{\mathbf{X}}_t, \gamma_{SR}\right)$ represents the PEP that $\mathbf{X}_t$ is transmitted at the SN and $\widetilde{\mathbf{X}}_t$ is detected at the RNs, and $P_{RD}\left(\widetilde{\mathbf{X}}_t \to \widehat{\mathbf{X}}_t, \gamma_{RD}\right)$ denotes the PEP that $\widetilde{\mathbf{X}}_t$ is transmitted at the RNs and $\widehat{\mathbf{X}}_t$ is detected at the DN. We note that because the DD algorithms directly retrieve the transmit block $\widetilde{\mathbf{S}}_t$, the evaluation of $\widetilde{\mathbf{X}}_t$ is based on the relationship of $\widetilde{\mathbf{S}}_t = \widetilde{\mathbf{S}}_{t-1}\widetilde{\mathbf{X}}_t$, which requires both the present and previous detected transmit blocks. However, when both $\widetilde{\mathbf{S}}_t$ and $\widetilde{\mathbf{S}}_{t-1}$ are illegal, the analysis might fall into severe trouble, because the derivation of $\widetilde{\mathbf{X}}_t$ from $\widetilde{\mathbf{S}}_t = \widetilde{\mathbf{S}}_{t-1}\widetilde{\mathbf{X}}_t$ may be a non-solution problem. Nevertheless, we could assume that the detection of the former transmitted transmit block $\widetilde{\mathbf{S}}_{t-1}$ is error-free, which means $\widetilde{\mathbf{S}}_{t-1} = \mathbf{S}_{t-1}$. Then, $\widetilde{\mathbf{X}}_t$ can be easily calculated by $\widetilde{\mathbf{X}}_t = \mathbf{S}_{t-1}^H \widetilde{\mathbf{S}}_t$.

Bearing this in mind, in the following, we evaluate $P_{RD}\left(\widetilde{\mathbf{X}}_t \to \widehat{\mathbf{X}}_t, \gamma_{RD}\right)$ first, which is suitable for both proposed DD algorithms at the RNs.

Lemma 5 *The PEP of the second hop with DD at the RNs is given by*

$$\begin{aligned} &P_{RD}\left(\widetilde{\mathbf{X}}_t \to \widehat{\mathbf{X}}_t, \gamma_{RD} | \mathbf{H}_{RD}, \mathbf{X}_t, \mathbf{S}_{t-1}\right) \\ &= Q\left(\sqrt{\frac{\gamma_{RD}}{2}} \frac{\|\mathbf{H}_{RD}\mathbf{S}_{t-1}\mathbf{C}_1\|^2 - \|\mathbf{H}_{RD}\mathbf{S}_{t-1}\mathbf{C}_2\|^2}{\sqrt{\|\mathbf{H}_{RD}\mathbf{S}_{t-1}\mathbf{C}\|^2 + \left\|\mathbf{H}_{RD}\mathbf{S}_{t-1}\widetilde{\mathbf{X}}_t\mathbf{C}^H\right\|^2}}\right), \end{aligned} \quad (3.103)$$

where $\mathbf{C}_1 = \widetilde{\mathbf{X}}_t - \mathbf{X}_t$, $\mathbf{C}_2 = \widetilde{\mathbf{X}}_t - \widehat{\mathbf{X}}_t$ *and* $\mathbf{C} = \mathbf{C}_2 - \mathbf{C}_1 = \mathbf{X}_t - \widehat{\mathbf{X}}_t$.

Proof From (3.97), we notice that the probability of the erroneous detection of $\widehat{\mathbf{X}}_t$ at the DN when $\mathbf{X}_t$ is transmitted at the SN, i.e., $P_{RD}\left(\widetilde{\mathbf{X}}_t \to \widehat{\mathbf{X}}_t, \gamma_{RD}\right)$, is equal to $\Pr(\left\|\mathbf{Y}_{D_t} - \mathbf{Y}_{D_{t-1}}\mathbf{X}_t\right\|^2 > \left\|\mathbf{Y}_{D_t} - \mathbf{Y}_{D_{t-1}}\hat{\mathbf{X}}_t\right\|^2)$, given that $\widetilde{\mathbf{X}}_t$ is detected at the RNs. By substituting (3.96) into the above inequation and recalling the assumption of $\widetilde{\mathbf{S}}_{t-1} = \mathbf{S}_{t-1}$, the ignorance of the second order noise terms results in

$$\begin{aligned} &P_{RD}\left(\widetilde{\mathbf{X}}_t \to \widehat{\mathbf{X}}_t, \gamma_{RD} | \mathbf{H}_{RD}, \mathbf{X}_t, \mathbf{S}_{t-1}\right) \\ &= \Pr\left(\left\|\mathbf{H}_{RD}\mathbf{S}_{t-1}\mathbf{C}_1 + \mathbf{W}_{D_t} - \mathbf{W}_{D_{t-1}}\mathbf{X}_t\right\|^2 \right. \\ &\qquad \left. > \left\|\mathbf{H}_{RD}\mathbf{S}_{t-1}\mathbf{C}_2 + \mathbf{W}_{D_t} - \mathbf{W}_{D_{t-1}}\widehat{\mathbf{X}}_t\right\|^2\right) \\ &= \Pr\left(\|\mathbf{H}_{RD}\mathbf{S}_{t-1}\mathbf{C}_1\|^2 - \|\mathbf{H}_{RD}\mathbf{S}_{t-1}\mathbf{C}_2\|^2 \right. \\ &\qquad \left. > 2\Re\left\{\mathrm{Tr}\left\{\begin{array}{r} (\mathbf{H}_{RD}\mathbf{S}_{t-1}\mathbf{C})^H\mathbf{W}_{D_t} + \\ \mathbf{H}_{RD}\mathbf{S}_{t-1}\widetilde{\mathbf{X}}_t\mathbf{C}^H\mathbf{W}_{D_{t-1}}^H + \mathbf{C}^H\mathbf{W}_{D_{t-1}}^H\mathbf{W}_{D_t} \end{array}\right\}\right\}\right) \\ &\approx \Pr\left(\|\mathbf{H}_{RD}\mathbf{S}_{t-1}\mathbf{C}_1\|^2 - \|\mathbf{H}_{RD}\mathbf{S}_{t-1}\mathbf{C}_2\|^2\right. \end{aligned}$$

$$> 2\Re\left\{\mathrm{Tr}\left\{(\mathbf{H}_{RD}\mathbf{S}_{t-1}\mathbf{C})^H\mathbf{W}_{D_t}+\mathbf{H}_{RD}\mathbf{S}_{t-1}\widetilde{\mathbf{X}}_t\mathbf{C}^H\mathbf{W}_{D_{t-1}}^H\right\}\right\}. \quad (3.104)$$

Taking note that

$$\begin{aligned}&\mathrm{Tr}\left\{(\mathbf{H}_{RD}\mathbf{S}_{t-1}\mathbf{C})^H\mathbf{W}_{D_t}+\mathbf{H}_{RD}\mathbf{S}_{t-1}\widetilde{\mathbf{X}}_t\mathbf{C}^H\mathbf{W}_{D_{t-1}}^H\right\}\\&\sim \mathcal{CN}\left(0,\left(\|\mathbf{H}_{RD}\mathbf{S}_{t-1}\mathbf{C}\|^2+\left\|\mathbf{H}_{RD}\mathbf{S}_{t-1}\widetilde{\mathbf{X}}_t\mathbf{C}^H\right\|^2\right)\Big/\gamma_{RD}\right),\end{aligned} \quad (3.105)$$

Equation (3.103) can be readily obtained. □

To arrive at a closed-form expression of (3.103), we have to average the Q-function over all channel realizations. However, The term inside the Q-function is composed of four correlated gamma RVs, which makes the evaluation a tough task. Thus, we decide to calculate the PEP of the second hop in a numerical way.

The next step to obtain (3.102) is to evaluate $P_{SR}\left(\mathbf{X}_t \to \widetilde{\mathbf{X}}_t, \gamma_{SR}\right)$. In order to simplify the analysis, we write it in another form as

$$\begin{aligned}P_{SR}\left(\mathbf{X}_t \to \widetilde{\mathbf{X}}_t, \gamma_{SR}\right) &= P_{SR}\left(\mathbf{S}_t \to \widetilde{\mathbf{S}}_t, \gamma_{SR}|\mathbf{S}_{t-1}\right)\\&=\prod_{i=1}^{N_R} E_{d_i}\left\{P_{SR_i}\left(\mathbf{s}_{t,i}\to\widetilde{\mathbf{s}}_{t,i},\gamma_{SR}|d_i,\mathbf{S}_{t-1}\right)\right\},\end{aligned} \quad (3.106)$$

where $P_{SR_i}\left(\mathbf{s}_{t,i}\to\widetilde{\mathbf{s}}_{t,i},\gamma_{SR}|d_i,\mathbf{S}_{t-1}\right)$ represents the conditional probability when $\mathbf{s}_{t,i}$ is transmitted to the i-th RN but $\widetilde{\mathbf{s}}_{t,i}$ is detected, given d_i and $\mathbf{S}_{t-1}$ and the corresponding $\mathbf{S}_t$ and $\widetilde{\mathbf{S}}_t$ can be obtained from $\mathbf{X}_t$ and $\widetilde{\mathbf{X}}_t$ with respect to $\mathbf{S}_{t-1}$. It is obvious that there are three possible error events between $\mathbf{s}_{t,i}$ and $\widetilde{\mathbf{s}}_{t,i}$, which can be defined as: (i) E1 $=\left\{\mathbf{e}_{j_i}\to\mathbf{e}_{j_i}, c_i\to\widetilde{c}_i\right\}$; (ii) E2 $=\left\{\mathbf{e}_{j_i}\to\mathbf{e}_{\widetilde{j}_i}, c_i\to c_i\right\}$; and (iii) E3 $=\left\{\mathbf{e}_{j_i}\to\mathbf{e}_{\widetilde{j}_i}, c_i\to\widetilde{c}_i\right\}$, which represent the signal error, the time-interval error and the joint signal and time-interval error, respectively.

Since two DD algorithms are proposed, in the following, we will complete the derivation for optimal ML detection and near-optimal detection respectively.

Lemma 6 *The conditional PEP of the first hop with optimal ML DD at the RNs is given by*

$$\begin{aligned}&P_{SR}\left(\mathbf{S}_t\to\widetilde{\mathbf{S}}_t,\gamma_{SR}\right)\\&=\prod_{i=1}^{N_R}\left\{\left(1-f'\left(\gamma_{SR},\vartheta_i\right)\right)^L\sum_{l=0}^{L-1}\frac{\mathrm{C}(L-1+l,l)}{2^{L+l}}\left(1+f'\left(\gamma_{SR},\vartheta_i\right)\right)^l\right\},\end{aligned} \quad (3.107)$$

where $f'\left(\gamma_{SR},\vartheta_i\right)=\sqrt{\vartheta_i\gamma_{SR}/(4+\vartheta_i\gamma_{RD})}$ *and* $\vartheta_i=\left\|\Delta\mathbf{s}_{t,i}\right\|^2=\left\|\mathbf{s}_{t,i}-\widetilde{\mathbf{s}}_{t,i}\right\|^2$.

Proof The PEP at the i-th RN under the optimal ML detection (3.93) can be expressed as

$$P_{SR_i}\left(\mathbf{s}_{t,i} \rightarrow \widetilde{\mathbf{s}}_{t,i}, \gamma_{SR}|d_i\right) = \Pr\left(\left\|\mathbf{y}_{R_t,i} - d_i\mathbf{s}_{t,i}\right\|^2 - \left\|\mathbf{y}_{R_t,i} - d_i\widetilde{\mathbf{s}}_{t,i}\right\|^2 > 0|d_i, \mathbf{S}_{t-1}\right), \tag{3.108}$$

where

$$\begin{aligned}
&\left\|\mathbf{y}_{R_t,i} - d_i\mathbf{s}_{t,i}\right\|^2 - \left\|\mathbf{y}_{R_t,i} - d_i\widetilde{\mathbf{s}}_{t,i}\right\|^2 \\
&= -d_i^2\left(\left|\mathbf{s}_{t,i}\right|^2 + \left|\widetilde{\mathbf{s}}_{t,i}\right|^2 - \mathbf{s}_{t,i}\widetilde{\mathbf{s}}_{t,i}^H - \mathbf{s}_{t,i}^H\widetilde{\mathbf{s}}_{t,i}\right) - 2d_i\Re\left\{\mathbf{w}_{R_t,i}\left(\mathbf{s}_{t,i} - \widetilde{\mathbf{s}}_{t,i}\right)^H\right\} \\
&= -d_i^2\vartheta_i - 2d_i\Re\left\{\mathbf{w}_{R_t,i}\Delta\mathbf{s}_{t,i}{}^H\right\}.
\end{aligned} \tag{3.109}$$

It's obvious that (3.109) conditioned on the channel realization is a complex-valued RV with mean $-d_i^2\vartheta_i$ and variance $4d_i^2\vartheta_i \big/ \gamma_{SR}$. Therefore, (3.108) can be written in terms of Q-function as

$$P_{SR_i}\left(\mathbf{s}_{t,i} \rightarrow \widetilde{\mathbf{s}}_{t,i}, \gamma_{SR}|d_i\right) = Q\left(\sqrt{\frac{\gamma_{SR}d_i^2\vartheta_i}{2}}\right). \tag{3.110}$$

Depending on the error events, ϑ_i in (3.110) takes the following values of:

$$\vartheta_i = \begin{cases} |c_i - \widetilde{c}_i|^2, & \text{E1} \\ 2 \quad , & \text{E2, E3} \end{cases}. \tag{3.111}$$

Then, by virtue of [[4], Eqs. (9.5) and (9.6)], we have

$$\begin{aligned}
&P_{SR_i}\left(\mathbf{s}_{t,i} \rightarrow \tilde{\mathbf{s}}_{t,i}, \gamma_{SR}\right) \\
&= \left(1 - f'\left(\gamma_{SR}, \vartheta_i\right)\right)^L \sum_{l=0}^{L-1} \frac{1}{2^{L+l}} \mathrm{C}(L - 1 + l, l)\left(1 + f'\left(\gamma_{SR}, \vartheta_i\right)\right)^l.
\end{aligned} \tag{3.112}$$

By substituting (3.112) into (3.106), we can complete the proof of the PEP (3.107).
□

Finally, substituting (3.103) and (3.107) into (3.102), and averaging it over $\widetilde{\mathbf{X}}_t$ and $\mathbf{S}_{t-1}$, we can obtain the ABEP upper bound (3.101) when applying optimal ML DD at the RNs.

Lemma 7 *The conditional PEP of the first hop with near-optimal DD at the RNs is given by*

$$P_{SR}\left(\mathbf{S}_t \rightarrow \widetilde{\mathbf{S}}_t, \gamma_{SR}\right) = \prod_{i=1}^{N_R} \{g\left(\gamma_{SR}, \vartheta_i\right)\}, \tag{3.113}$$

with

$$g\left(\gamma_{SR},\vartheta_i\right)=\begin{cases}\left(1-f'\left(\gamma_{SR},\vartheta_i\right)\right)^L\sum\limits_{l=0}^{L-1}\frac{C(L-1+l,l)}{2^{L+l}}\left(1+f'\left(\gamma_{SR},\vartheta_i\right)\right)^l, & \text{E1},\\ \frac{2^{1-L}(2+\gamma_{SR})^{-L}}{M} & \text{E2, E3}\end{cases}$$

Proof Following the definition of the three error events, we can easily see from (3.95) that the conditional PEP of E1 at the j-th RN is exactly the same as that in (3.112).

When E2 or E3 happens at the i-th RN, wrong time interval decision occurs with the probability of

$$\begin{aligned}P_{SR_i}\left(\mathbf{e}_{j_i}\rightarrow\mathbf{e}_{\bar{j}_i},\gamma_{SR}|d_i\right)&=\Pr\left(\left|\mathbf{y}_{R_t,i}\mathbf{e}_{j_i}^T\right|<\left|\mathbf{y}_{R_t,i}\mathbf{e}_{\bar{j}_i}^T\right||d_i\right)\\&=\Pr\left(\left|d_ic_i+\mathbf{w}_{R_t,i}\mathbf{e}_{j_i}^T\right|<\left|\mathbf{w}_{R_t,i}\mathbf{e}_{\bar{j}_i}^T\right||d_i\right).\end{aligned}\tag{3.114}$$

It's obvious that $\left|d_ic_i+\mathbf{w}_{R_t,i}\mathbf{e}_{j_i}^T\right|$ follows the Rice distribution of

$$f\left(x,\gamma_{SR}\right)=2\gamma_{SR}x\exp\left(-\left(x^2+d_i^2\right)\gamma_{SR}\right)I_0\left(2d_j\gamma_{SR}x\right),\tag{3.115}$$

where $I_0\left(\cdot\right)$ represents the zero-order modified bessel function of the first kind [[5], p. 911]. Meanwhile, $\left|\mathbf{w}_{R_t,i}\mathbf{e}_{\bar{j}_i}^T\right|$ follows the Rayleigh distribution of

$$f\left(x,\gamma_{SR}\right)=2\gamma_{SR}x\exp\left(-x^2\gamma_{SR}\right).\tag{3.116}$$

Therefore, (3.114) can be straightforwardly calculated as

$$P_{SR_i}\left(\mathbf{e}_{j_i}\rightarrow\mathbf{e}_{\bar{j}_i},\gamma_{SR}|d_i\right)=\frac{1}{2}e^{-\frac{d_i^2\gamma_{SR}}{2}}.\tag{3.117}$$

By averaging (3.117) over d_i^2, we obtain

$$P_{SR_i}\left(\mathbf{e}_{j_i}\rightarrow\mathbf{e}_{\bar{j}_i},\gamma_{SR}\right)=2^{1-L}(2+\gamma_{SR})^{-L}.\tag{3.118}$$

When the erroneous detection of the time interval happens, the symbol detection depends randomly on the AWGN. Since both the M-PSK constellation and the AWGN are circularly symmetric, the probabilities of E2 and E3 are both $1/M$ of (3.118), i.e.,

$$\begin{aligned}P_{SR_i}\left(\mathbf{e}_{j_i}\rightarrow\mathbf{e}_{\bar{j}_i},c_i\rightarrow c_i,\gamma_{SR}\right)&=P_{SR_i}\left(\mathbf{e}_{j_i}\rightarrow\mathbf{e}_{\bar{j}_i},c_i\rightarrow\widetilde{c}_i,\gamma_{SR}\right)\\&=\frac{2^{1-L}(2+\gamma_{SR})^{-L}}{M}.\end{aligned}\tag{3.119}$$

Finally, substituting (3.112) and (3.119) into (3.106) concludes the proof of (3.113). □

The ABEP upper bound (3.101) when applying near-optimal DD at the RNs can be readily obtained by plugging (3.103) and (3.113) into (3.102), and then averaging it over $\widetilde{\mathbf{X}}_t$ and $\mathbf{S}_{t-1}$.

3.4.3 Performance Evaluation

Monte Carlo simulations are conducted to examine the performance of the proposed DH-HDSM and verify the theoretical analysis.

For comparison, the dual-hop hybrid spatial modulation (DH-HSM) architecture in [12] is introduced as a benchmark. Besides, since the proposed low complexity DH-HDSM architecture mainly benefits from differential modulation, classical VMIMO architecture with DPSK modulation is also considered. Since each RN is active only once during each transmit block, for fairness of the comparison, we consider the single-RN-activation systems of (i) single relay system and (ii) the best relay selection system. In the differential VMIMO architectures, the same linear pre-coding methods are used at the SN, and the DN detects the received signal via the differential ML detector.

In the single relay system, which will be referred to as dual-hop DPSK with 1 relay (DH-Dw1R), the SN conveys the DPSK symbol to the single RN. We note that the detection method to obtain the constellation symbols in (3.95) can be applied to achieve optimal performance and minimum complexity. Besides, when the same symbol constellation size is utilized, the proposed detection method in DH-Dw1R results in the same computational complexity as the near-optimal DD algorithm in DH-HDSM. Recalling the fact that DD is applied once per transmit block in DH-HDSM at every RN, while the single RN in DH-Dw1R detects the symbol every time it receives a symbol, we notice that the overall complexity at the RNs in DH-HDSM is the same as DH-Dw1R. In other words, the computation load is distributed to the multiple RNs by DH-HDSM. Regarding the EE, since each RN transmits a unit power symbol every $2N_R$ time slots in DH-HDSM, the overall transmit power consumption at the RNs is also identical to that in DH-Dw1R. However, since additional space bits can be carried by the patterns of the information blocks, we could draw the conclusion that DH-HDSM achieves improved spectrum efficiency than DH-Dw1R.

In the following, the best relay selection system will be denoted as dual-hop DPSK with best relay selection (DH-DwBRS). In DH-DwBRS, the SN will select the RN with the most preferable RN-DN channel gain and send DPSK symbols via pre-coding. The selected RN employs the same detection method as in DH-Dw1R and transmits the detected symbol to the DN. We note that the comparison between DH-HDSM and DH-DwBRS is unfair, because the selection of the best RN at the SN requires the knowledge of the instantaneous RNs-DN link CSI, which means feedback links have to be placed in both hops. However, we show that the proposed DH-HDSM still has enhanced BER performance in the majority of the considered system setups even with a simplified open-loop architecture. Also, apart form the feedback links to obtain the CSI between RNs and DN, the detection complexity of DH-DwBRS at the RNs is the same as in DH-Dw1R.

For simplicity and without loss of generality, we assume the SNR during the first hop and the second hop is the same, i.e., $\sigma_R^2 = \sigma_D^2 = N_0$, and the overall SNR is defined as $\gamma = (P_s + P_r)/\sigma^2 = 2\gamma_{SR} = 2\gamma_{RD}$ throughout the simulations.

Performance Comparison Between DH-HDSM and DH-HSM

Figures 3.15 and 3.16 compare the BER performance of DH-HDSM and DH-HSM with $N_S = 8$, $N_R = 2$ at the data rate of 1 bps/Hz, where DH-HSM applies BPSK constellation and DH-HDSM uses BPSK and QPSK for different non-zero elements to achieve the same data rate. Both ZF and MMSE pre-coding are considered in the simulations.

In Fig. 3.15, the CD algorithm is considered at the RNs. As shown in the figure, due to the application of differential modulation, an about 2 dB BER performance loss of the proposed DH-HDSM can be observed as compared with DH-HSM. Besides, for a small number of receive antennas $N_D = \{1, 2, 4\}$, MMSE and ZF pre-coding have similar performance, because as can be seen from the performance analysis, the error performance is dominated by the hop with lower diversity order, i.e., the second hop. When $N_D = 16$, MMSE achieves an about 0.5 dB gain from the minimized mean square error (MSE).

Next, Fig. 3.16 considers the scenario where DD is performed at the RNs. For simplicity, we take the near-optimal algorithm as the representative for DH-HDSM. Besides, simulation curves of CD algorithm with ZF pre-coding are also plotted for reference to reveal the performance penalty introduced by DD. As shown in the figure, for $N_D = 16$, DH-HDSM outperforms DH-HSM for about 2 dB when DD is applied at the RNs. Also, the performance difference between the CD and DD algorithms is indistinguishable for DH-HDSM except for $N_D = 16$ in very low SNRs. However, for DH-HSM, as N_D increases to 4, although the curves merge at

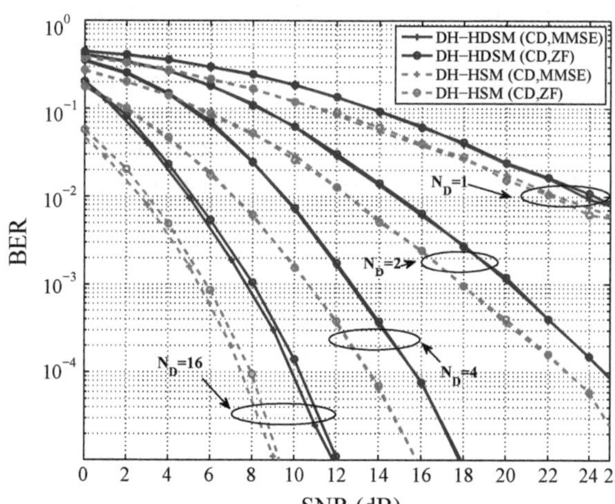

Fig. 3.15 BER performance of DH-HDSM and DH-HSM with the system setup of $N_S = 8$, $N_R = 2$ and $N_D = \{1, 2, 4, 16\}$ at a data rate of $k = 1$ bps/Hz when CD is applied at the RNs

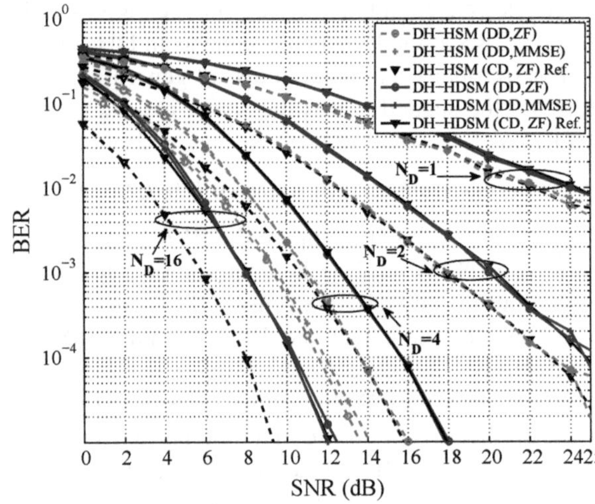

Fig. 3.16 BER performance of DH-HDSM and DH-HSM with the system setup of $N_S = 8$, $N_R = 2$ and $N_D = \{1, 2, 4, 16\}$ at a data rate of $k = 1$ bps/Hz when DD is applied at the RNs

high SNR, the performance gap between CD and DD at the BER of 1×10^{-2} is up to 1 dB. When N_D goes up to 16, the performance difference increases to about 4 dB. This is because when the receive antenna number is small, the error occurs mainly in the second hop. Thus, different detection strategies at the RNs affect little of the performance. However, as the receive antenna number increases, due to the improved diversity order, the detection of the second hop becomes nearly error-free even at very low SNR. Thus, the performance is mainly dominated by the error detection at the RNs. Since the DD algorithm in DH-HDSM exploits the time domain properties of the transmit blocks and performs joint detection, the performance loss compared with optimal ML CD algorithm is negligible. However, the DD of the DH-HSM detects the symbol one by one, which will easily cause the multiple relay activation in the second hop, and result in the hug performance gap with CD. In conclusion, the performance gain of DH-HDSM in comparison with DH-HSM comes from the moderate performance penalty as compared with CD.

Moreover, in Fig. 3.16, the comparison between MMSE and ZF pre-coding reveals that while maintaining similar performance to ZF in DH-HDSM, MMSE pre-coding suffers an about 0.5 dB loss of SNR in DH-HSM when $N_D = 16$. This is because MMSE pre-coding introduces interference among RNs, while CD conducts joint detection and can benefit from the minimized symbol MSE, the performance of the DD algorithm suffers from the introduced interference because the detection is done at the RNs independently. For DH-HDSM, since the interference remains constant at a specific RN for a given SNR, according to (3.94) and (3.95), it doesn't affect the detection of time intervals as compared with ZF pre-coding, which results in, as shown in Fig. 3.16, negligible performance loss. However, the interference will affect the detection of activation states in DH-HSM and cause larger probability of multiple active RNs during the second hop. At a large N_D, since the detection

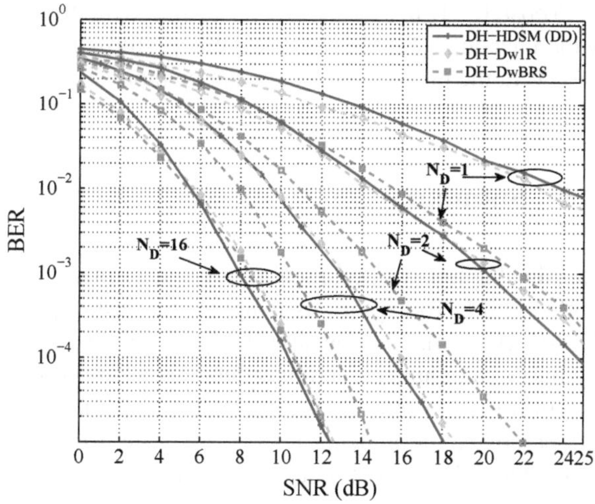

Fig. 3.17 BER performance of DH-HDSM and other differential VMIMO architectures of DH-Dw1R and DH=DwBRS with the system setup of $N_S = 8, N_R = 2$ and $N_D = \{1, 2, 4, 16\}$ at a data rate of $k = 1$ bps/Hz

at the RNs becomes the bottleneck of the system, the performance degradation is noticeable.

Performance Comparison Between DH-HDSM and Other Differential VMIMO Architectures

In Fig. 3.17, we evaluate the performance of DH-HDSM with the differential VMIMO architectures of DH-Dw1R and DH-DwBRS. Near-optimal DD is considered for DH-HDSM to achieve minimum complexity. As shown in the figure, when $N_D = 1$, DH-Dw1R slightly outperforms DH-HDSM. As N_D increases to more than 2, DH-HDSM demonstrates an about 0.5 dB performance gain against DH-Dw1R in the high SNR region. This can be explained by the resolution of the space bits at the DN. To be more specific, when N_D is small, the detection of the space bits, i.e., the RN activation sequence, is highly dependent on the instantaneous channel gains between each RN and the DN antennas. Thus, error detection may easily occur when the channels are poor. When that happens, the detection of the constellation bits is randomly decided by the noise and results in performance degradation. When N_D is large, due to the increased diversity order, the resolution of the space bits increases, and the error detection is mainly within the constellation bits. Thanks to the additional space bits, DH-HDSM can use a smaller symbol constellation as compared with DH-Dw1R to reach the target data rate, and thus can offer better performance. As for DH-DwBRS, although the comparison is unfair, we notice that when N_D is large, DH-HDSM offers better performance in the considered SNR region. However, due to the higher diversity order provided by relay selection, DH-DwBRS will offer better performance at higher SNR. We note that the performance gain of DH-HDSM against DH-Dw1R and DH-DwBRS is mainly provided by the multiplexing gain from the increased space bits. In the considered scenario, since only 2 RNs help to form the DH-HDSM architecture, the data rate provided by the space bits is only

0.25 bps/Hz. Thus, the reduction of the symbol constellation size is insignificant, which accounts for the inappreciable performance difference. When more RNs are involved, the performance gain can be more significant.

3.5 Summary

In this chapter, we designed a new space-time domain index modulation scheme, called DSM. To explore its insights, we further provided the design guidelines for DSM, which mainly consider the low-complexity detector and application of Gray coding in antenna activation orders. We also proposed the receive DSM that maps the information into the receive antennas. Finally, we applied DSM and PDSM to cooperative communication scenarios, where a DH-HDSM architecture is designed. In the DH-HDSM architecture, we further proposed two different detection protocols CD and DD at the RNs, both of which achieve reduced complexity at the RNs and DN. The performance analyses of the above schemes were well investigated.

References

1. V. Tarokh, H. Jafarkhani, A differential detection scheme for transmit diversity. IEEE J. Sel. Areas Commun. **18**(7), 1169–1174 (2000)
2. http://en.wikipedia.org/wiki/Stirling's_approximation
3. http://en.wikipedia.org/wiki/Factorial_number_system
4. M.K. Simon, M.S. Alouini, *Digital Communication Over Fading Channels: A Unified Approach to Performance Analysis*, 2nd edn. (Wiley, New York, 2004)
5. I.S. Gradshteyn, I.M. Rhyhik, *Table of Integrals, Series and Products*, 7th edn. (Academic, New York, 2007)
6. H. Jafarkhani, A quasi-orthogonal space-time block code. IEEE Trans. Commun. **49**(1), 1–4 (2001)
7. R. Mesleh, H. Haas, S. Sinanovic, C. Ahn, S. Yun, Spatial modulation. IEEE Trans. Veh. Technol. **57**(4), 2228–2242 (2008)
8. C. Xu, S. Sugiura, S.X. Ng, L. Hanzo, Spatial modulation and space-time shift keying: optimal performance at a reduced detection complexity. IEEE Trans. Commun. **61**, 206–216 (2013)
9. A.J. Viterbi, Error bounds for convolutional codes and an asymptotically optimum decoding algorithm. IEEE Trans. Inf. Theory **IT–13**, 260–269 (1967)
10. D.L. Kreher, D.R. Stinson, *Combinatorial Algorithms: Generation, Enumeration, and Search (Discrete Mathematics and Its Applications)*, 1st edn. (CRC Press, Boca Raton, 1998)
11. N. Balakrishnan, C.D. Lai, *Continuous Bivariate Distributions* (Springer, New York, 2009)
12. A. Stavridis, D. Basnayaka, S. Sinanovic, M. Di Renzo, H. Haas, A virtual MIMO dual-hop architecture based on hybrid spatial modulation. IEEE Trans. Commun. **62**(9), 3161–3179 (2014)
13. B. Hochwald, S. Vishwanath, Space-time multiple access: linear growth in the sum rate, in *Proceedings of the 40th Annual Allerton Conference on Communications, Control and Computing, Monticello, IL* (2002)

Chapter 4
Frequency Domain Index Modulation

Chapters 2 and 3 discuss how index modulation operates in the space and space-time domains, respectively. In this chapter, we focus on the implementation of index modulation in the frequency domain. First, the orthogonal frequency division multiplexing with index modulation (OFDM-IM), which is a representative of this sort, is introduced. Then, design guidelines for improving the diversity gain, coding gain, and spectral efficiency of OFDM-IM systems are provided. Finally, we present the idea of integrating intercarrier interference (ICI) self-cancellation techniques into the OFDM-IM framework to combat the severe ICI in the Vehicle to X (V2X) communications.

4.1 OFDM with Index Modulation

In this section, we introduce OFDM-IM, which conveys additional information through the indices of OFDM active subcarriers. The achievable rate of OFDM-IM is also analyzed.

4.1.1 Transceiver Structure

Suppose that the system bandwidth is B_T, which is occupied by a total of N OFDM subcarriers, i.e., subcarrier spacing is $B_S = B_T/N$. In OFDM-IM, the N subcarriers are split into $G = N/L$ groups, each of which consists of L subcarriers. For group g, the subcarrier indices are given by

$$\Psi^g = \left\{\beta_1^g, \ldots, \beta_L^g\right\} \tag{4.1}$$

where $\beta_l^g \in \{1, \ldots, N\}$ with $g = 1, \ldots, G$ and $l = 1, \ldots, L$. Note that $\Psi^{g_1} \cap \Psi^{g_2} = \emptyset$ and $\Psi^1 \cup \cdots \cup \Psi^G = \{1, \ldots, N\}$, where $g_1, g_2 \in \{1, \ldots, G\}$ and $g_1 \neq g_2$.

M. Wen et al., *Index Modulation for 5G Wireless Communications*,
Wireless Networks, DOI 10.1007/978-3-319-51355-3_4

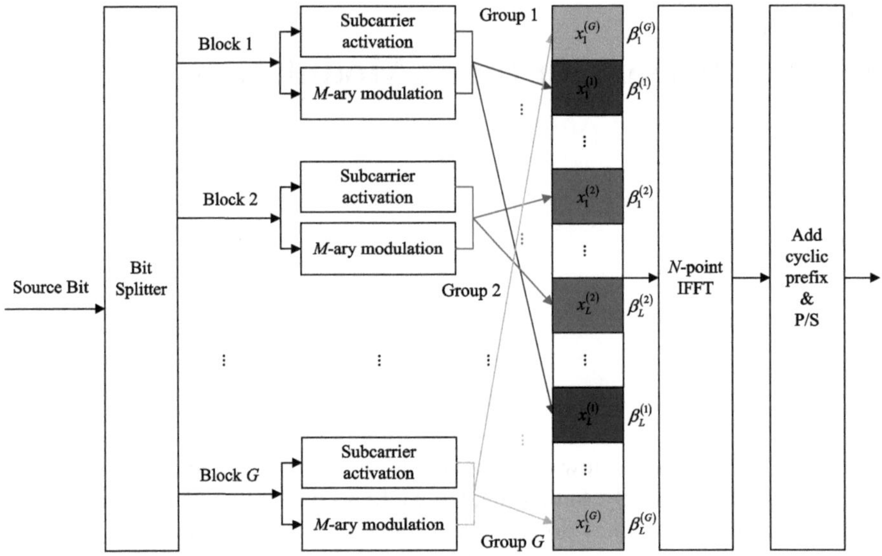

Fig. 4.1 Baseband transmitter structure of OFDM-IM with arbitrary grouping

It is assumed that both the transmitter and receiver have agreed on the subcarrier grouping in (4.1). The reason why subcarrier grouping is employed in OFDM-IM will be clarified in the sequel.

Figure 4.1 depicts the OFDM-IM baseband transmitter structure. The source bits are equally split into G blocks, each of which consists of two parts. The bits associated with bit block g determine the states of all subcarriers and modulate the symbols carried on the active subcarriers within subcarrier group g. Specifically, m out of L subcarriers within the subcarrier group are set to be inactive according to the bits of the first part and the remaining $L - m$ subcarriers are activated to transmit $(L - m)$ M-ary PSK/QAM modulated symbols according to the bits of the second part. Note that given L and m, there are in total $\mathrm{C}(L, m)$ combinations of active/inactive subcarrier indices (or say subcarrier activation patterns); however, to transmit integer number of bits only $2^{\lfloor \log_2(\mathrm{C}(L,m)) \rfloor}$ ones are utilized though this is not limited in this Section. Take $L = 4$ and $m = 2$ for example. All $\mathrm{C}(4, 2) = 6$ combinations for subcarrier group g are listed as follows:

$$\Omega_1^g = \{\beta_1^g, \beta_2^g\} \left(\bar{\Omega}_1^g = \{\beta_3^g, \beta_4^g\}\right), \ \Omega_2^g = \{\beta_1^g, \beta_3^g\} \left(\bar{\Omega}_2^g = \{\beta_2^g, \beta_4^g\}\right),$$
$$\Omega_3^g = \{\beta_1^g, \beta_4^g\} \left(\bar{\Omega}_3^g = \{\beta_2^g, \beta_3^g\}\right), \ \Omega_4^g = \{\beta_2^g, \beta_3^g\} \left(\bar{\Omega}_4^g = \{\beta_1^g, \beta_4^g\}\right),$$
$$\Omega_5^g = \{\beta_2^g, \beta_4^g\} \left(\bar{\Omega}_5^g = \{\beta_1^g, \beta_3^g\}\right), \ \Omega_6^g = \{\beta_3^g, \beta_4^g\} \left(\bar{\Omega}_6^g = \{\beta_1^g, \beta_2^g\}\right),$$

where Ω_j^g represents the j-th combination of the active subcarrier indices for subcarrier group g and $\bar{\Omega}_j^g$ is the complement of Ω_j^g with respect to the set Ψ^g, which therefore denotes the j-th combination of the inactive subcarrier indices for subcarrier group g. The way to select one out of $\mathrm{C}(L, m)$ combinations of active/insubcarrier indices according to random bits will be discussed in Sect. 4.2.2. Assume that the output of the index selector for active subcarriers is

$$\mathbf{x}_c^g \in \left\{\Omega_1^g, \ldots, \Omega_{\mathrm{C}(L,m)}^g\right\}. \tag{4.2}$$

At this point, it is clear that subcarrier grouping is necessary for OFDM-IM since without it we have $L = N$ and $\mathrm{C}(N, m)$ can take very large values, which makes the implementation complexity of OFDM-IM high. On the other hand, assume that the vector of the modulated symbols at the output of the M-ary modulator is given by

$$\mathbf{x}_s^g \in \mathcal{X}^{L-m}. \tag{4.3}$$

Denote the total transmit power of the system as P_T. By taking into account $\mathbf{x}_c^g$ and $\mathbf{x}_s^g$ for all g and concatenating G subcarrier groups, the frequency-domain OFDM symbol is created as

$$\mathbf{x} = [x_1, \ldots, x_N]^T, \tag{4.4}$$

where $x_i \in \{0, \sqrt{P_G/(L-m)}\mathbf{x}_s^1, \ldots, \sqrt{P_G/(L-m)}\mathbf{x}_s^G\}$ with $P_G = P_T/G$ representing the total transmit power within each subcarrier group and $i = 1, \ldots, N$. Note that as the transmitter has no knowledge of the CSI, it is wise to allocate active subcarriers with equal power. Also note that the power allocated to each active subcarrier is $P_G/(L-m)$ for OFDM-IM rather than P_G/L for classical OFDM in order to balance the total transmit power. Alternatively, it is feasible to constrain the transmit power of each active subcarrier to be the same as that of classical OFDM. However, though the latter policy has the merit of lower peak to average power ratio (PAPR), it is unfavorable for OFDM-IM in the maximization of information transmission capability. Before transmission, the inverse fast Fourier transform (IFFT) is applied to (4.4), yielding

$$\breve{x}_k = \frac{1}{\sqrt{N}} \sum_{i=1}^{N} x_i e^{\frac{2\sqrt{-1}\pi(i-1)(k-1)}{N}}, \quad k = 1, \ldots, N, \tag{4.5}$$

and a length-Q cyclic prefix (CP) of samples $\left[\breve{x}_{N-Q+1}, \ldots, \breve{x}_N\right]^T$ is appended to the beginning of the time-domain main OFDM symbol $\breve{\mathbf{x}} = \left[\breve{x}_1, \ldots, \breve{x}_N\right]^T$.

Consider a slowly time-varying multipath Rayleigh fading channel with channel impulse response

$$\breve{\mathbf{h}} = \left[\breve{h}_1, \ldots, \breve{h}_D\right]^T, \tag{4.6}$$

where D is the number of channel taps, and $\breve{h}_d$ $(d = 1, \ldots, D)$ is circularly symmetric complex Gaussian distributed with $E_{\breve{h}_d}[|\breve{h}_d|^2] = \delta_d^2$, and $E_{\breve{\mathbf{h}}}[\breve{\mathbf{h}}^H \breve{\mathbf{h}}] = \sum_{d=1}^{D} \delta_d^2 = 1$. Suppose that the CP length Q is no smaller than $D - 1$. After the removal of the CP of the received signal and the application of the FFT, the frequency-domain received signal on the i-th subcarrier is given by

$$y_i = h_i x_i + w_i, \quad i = 1, \ldots, N, \tag{4.7}$$

where $h_i = \sum_{d=1}^{D} \breve{h}_d e^{-2\sqrt{-1}\pi(i-1)(d-1)/N}$ is the channel coefficient and w_i is AWGN of power σ^2 at the i-th subcarrier, whose vector representations are given as $\mathbf{h}$ and $\mathbf{w}$, respectively. Note that $E_{\mathbf{h}}\left[\mathbf{h}^H \mathbf{h}\right] = N$ and $\sigma^2 = N_0 B_S$, where N_0 is the power spectral density of the AWGN. The task of the OFDM-IM receiver is to jointly detect the active/inactive subcarrier indices and the modulated symbols for each subcarrier group from (4.7). Assume that the channel estimates, which can be obtained by transmitting pilots prior to data transmission, are perfect. For subcarrier group g, ML detection can be carried out via

$$\begin{aligned}\left(\hat{\mathbf{x}}_c^g, \hat{\mathbf{x}}_s^g\right) &= \arg\min_{\mathbf{x}_c^g, \mathbf{x}_s^g} \sum_{i \in \Psi^g} |y_i - h_i x_i|^2 \\ &= \arg\min_{\mathbf{x}_c^g, \mathbf{x}_s^g} \|\mathbf{y}^g - \mathrm{diag}\{\mathbf{h}^g\}\, \mathbf{x}^g\|^2,\end{aligned} \tag{4.8}$$

where $\mathbf{y}^g$, $\mathbf{h}^g$, and $\mathbf{x}^g$ are vectors with elements being the frequency-domain received signals, channel coefficients, and frequency-domain transmitted symbols for subcarrier group g, respectively.

Note that from the principle of OFDM-IM described above, one can discover that the index modulation always works efficiently even if all subcarriers are faded completely correlatively, i.e., $h_1 = \cdots = h_N$, thanks to the IFFT at the transmitter and the FFT at the receiver. This is quite different from the SM technique, which fails when the channels between different transmit antennas are highly correlated.

Since (4.8) has to search through $\mathrm{C}(L, m)$ combinations from (4.2) and $\mathcal{X}^{L-m}$ from (4.3), the computational complexity grows exponentially with $\mathrm{C}(L, m)$ and $L - m$ according to $\mathcal{O}\left(\mathrm{C}(L, m)\, M^{L-m}\right)$. In the sequel, we introduce two low-complexity OFDM-IM detectors, namely log-likelihood ratio (LLR) detector and two-step ML detector, respectively, which lead to similar performance to the optimal ML detector but with much lower computational complexity.

LLR Detector

The LLR detector operates by first calculating the LLR values associated with all subcarriers from

$$LLR_i = \log \frac{\Pr(\mathrm{A}_i | y_i)}{\Pr(\bar{\mathrm{A}}_i | y_i)} = \log \frac{\Pr(\mathrm{A}_i)}{\Pr(\bar{\mathrm{A}}_i)} + \log \frac{f(y_i | \mathrm{A}_i)}{f(y_i | \bar{\mathrm{A}}_i)} \tag{4.9}$$

with $i = 1, \ldots, N$, where A_i represents the event that the i-th subcarrier is active while $\bar{\mathrm{A}}_i$ means just the opposite, and

$$\log \frac{f(y_i|\mathrm{A}_i)}{f(y_i|\bar{\mathrm{A}}_i)} = \frac{|y_i|^2}{N_0} + \log \frac{1}{M} \sum_{\tilde{s} \in \mathcal{X}} \frac{|y_i - h_i \tilde{s}|^2}{N_0}. \tag{4.10}$$

Then, it sorts all LLR values associated with a subcarrier group in descending order and picks out the first $L - m$ ones, whose corresponding subcarriers are simply regarded to be active. Finally, the symbols carried on the determined active subcarriers are demodulated independently.

It is clear from (4.9) that the computational complexity of the LLR detector is about $\mathcal{O}(M)$ per subcarrier, which is the same as that of the classical OFDM detector.

Two-Step ML Detector

Inspiring by the fact that the transmitted modulated symbols are mutually independent, we propose to decompose the ML detector in (4.8) into two steps to reduce the computational complexity. In the first step, we demodulate the received signals at all subcarriers within subcarrier group g as follows:

$$\hat{s}_i = \arg\min_{s \in \mathcal{X}} \left| y_i - h_i \sqrt{\frac{P_G}{L - m}} s \right|^2, \quad i \in \Psi^g. \tag{4.11}$$

Note that the outputs of (4.11) for each subcarrier group are L symbol estimates, but only $L - m$ ones are necessary. In the second step, we first calculate the following ML values:

$$\eta_i = 2\Re \left\{ y_i^* h_i \sqrt{\frac{P_G}{L - m}} \hat{s}_i \right\} - \frac{P_G |h_i \hat{s}_i|^2}{L - m}, \quad i \in \Psi^g. \tag{4.12}$$

Then, we decide on one hand on the $L - m$ active indices that have maximum ML values in (4.12), obtaining $\hat{\mathbf{x}}_c^g$, and on the other hand on the $L - m$ symbol estimates in (4.11) with indices $\hat{\mathbf{x}}_c^g$, obtaining $\hat{\mathbf{x}}_s^g$. It is clear from (4.11) and (4.12) that the computational complexity of the two-step ML detector is also about $\mathcal{O}(M)$. However, unlike the LLR detector, the two-step ML detector dispenses with the knowledge of the noise power at the receiver. Finally, the receiver de-maps $\left(\hat{\mathbf{x}}_c^g, \hat{\mathbf{x}}_s^g\right)$ to the corresponding information bits in reverse operations to the index mapper and M-ary modulation for subcarrier group g, and collects all information bits from G subcarrier groups to complete reception of an OFDM symbol.

4.1.2 Achievable Rate Analysis

In this subsection, we investigate the (ergodic) achievable rate of OFDM-IM assuming CSI at the receiver (CSIR). For fair comparison between OFDM-IM and classical OFDM, we resort to a unified definition of SNR as $\gamma = P_T / N_0 B_T = P_G / L\sigma^2$. The achievable rate of OFDM-IM is defined as [1]

$$R_{\mathrm{OFDM-IM}}(\gamma) = \frac{1}{B_T}\left(B_S E_{\mathbf{h}}\left[\mathrm{I}(\mathbf{x};\mathbf{y}|\mathbf{h})\right]\right)$$
$$= \frac{1}{N} E_{\mathbf{h}}\left[\mathrm{I}(\mathbf{x};\mathbf{y}|\mathbf{h})\right], \tag{4.13}$$

where $\mathrm{I}(\cdot,\cdot)$ denotes the mutual information. Note that (4.13) is exactly the spectral efficiency of OFDM-IM, measured in bps/Hz [2]. For simplicity, we restrict our analysis to the OFDM-IM system with regular subcarrier grouping, which implies that the frequency correlation is the same for all subcarrier groups, though it can be easily extended to arbitrary subcarrier grouping. An example for regular subcarrier grouping is the existing localized grouping, since under this method the (a,b)-th entry of the covariance matrix of $\mathbf{h}^g$, defined as $\Sigma = E_{\mathbf{h}^g}[\mathbf{h}^g(\mathbf{h}^g)^H]$, equals $[\Sigma_{\mathrm{LG}}]_{a,b} = \sum_{d=1}^{D} \delta_d^2 e^{-2\sqrt{-1}\pi(a-b)(d-1)/N}$ with $a, b = 1, \ldots, L$, which does not depend on g. Now, under the assumption of regular subcarrier grouping and based on the fact that the encoding and decoding processes associated with each subcarrier group are identical and statistically independent, (4.13) can be simplified as

$$R_{\mathrm{OFDM-IM}}(\gamma) = \frac{1}{N}\sum_{g'=1}^{G} E_{\mathbf{h}^{g'}}\left[\mathrm{I}\left(\mathbf{x}^{g'};\mathbf{y}^{g'}|\mathbf{h}^{g'}\right)\right]$$
$$= \frac{1}{L} E_{\mathbf{h}^g}\left[\mathrm{I}\left(\mathbf{x}_s^g, \mathbf{x}_c^g; \mathbf{y}^g|\mathbf{h}^g\right)\right], \quad \forall g \in \{1,\ldots,G\}. \tag{4.14}$$

According to (4.14), in the sequel we will only focus on one subcarrier group for achievable rate analysis. For brevity, we omit the superscript g for all variables defined above. In addition, we let $\beta_l = l$ for notational simplicity such that all subcarriers within the subcarrier group will be indexed from $\{1,\ldots,L\}$. Note that with the above notation, we will refer to the i-th subcarrier simply as the β_i-th subcarrier of OFDM signals.

Finite Constellation Input

Let us consider $\mathbf{x}_s$ to be a finite signal set, i.e., M is finite. However, for practical interests, we restrict the signal set to be a discrete constellation, such as M-ary PSK or QAM, where the constellation points are equiprobable.

Proposition 1 *The achievable rate of OFDM-IM is given by*

$$R_{\mathrm{OFDM-IM}}(\gamma) = \frac{L-m}{L}\log_2(M) + \frac{\log_2(\mathrm{C}(L,m))}{L} - \left(\log_2(e) - 1\right) - \lambda(\gamma), \tag{4.15}$$

with

$$\lambda(\gamma) = \frac{M^{m-L}}{L\mathrm{C}(L,m)} \sum_{j=1}^{\mathrm{C}(L,m)} \sum_{p^{(L-m)}} E_{\mathbf{v},\mathbf{h}}\left[\log_2\left(\sum_{j'=1}^{\mathrm{C}(L,m)} \sum_{p'^{(L-m)}} \prod_{i=1}^{L} \xi(v_i, h_i, \gamma)\right)\right], \tag{4.16}$$

where

$$\xi(v_i, h_i, \gamma) = \begin{cases} 2e^{-\left|v_i + h_i\sqrt{\frac{\gamma L}{L-m}}\left(s_{p_{\Omega_j^{-1}(i)}} - s_{p'_{\Omega_{j'}^{-1}(i)}}\right)\right|^2}, & i \in \Omega_j \cap \Omega_{j'} \\ 2e^{-\left|v_i + h_i\sqrt{\frac{\gamma L}{L-m}} s_{p_{\Omega_j^{-1}(i)}}\right|^2}, & i \in \Omega_j \cap \bar{\Omega}_{j'} \\ 2e^{-\left|v_i - h_i\sqrt{\frac{\gamma L}{L-m}} s_{p'_{\Omega_{j'}^{-1}(i)}}\right|^2}, & i \in \bar{\Omega}_j \cap \Omega_{j'} \\ 2e^{-|v_i|^2}, & i \in \bar{\Omega}_j \cap \overline{\Omega}_{j'} \end{cases}, \quad (4.17)$$

and $\mathbf{v} = [v_1, \ldots, v_L]^T$ *is a complex Gaussian random vector with zero mean and covariance matrix* $\mathbf{I}_L$.

Proof From (4.14), the achievable rate of OFDM-IM is given by

$$R_{\text{OFDM-IM}}(\gamma) = \frac{1}{L}\left[\mathrm{H}(\mathbf{x}_s, \mathbf{x}_c) - E_{\mathbf{h}}\left[\mathrm{H}(\mathbf{x}_s, \mathbf{x}_c|\mathbf{y}, \mathbf{h})\right]\right] \quad (4.18a)$$

$$= \frac{L-m}{L}\log_2(M) + \frac{\log_2(\mathrm{C}(L,m))}{L} - \frac{1}{L}E_{\mathbf{h}}\left[\mathrm{H}(\mathbf{x}_s, \mathbf{x}_c|\mathbf{y}, \mathbf{h})\right], \quad (4.18b)$$

where $\mathrm{H}(\cdot)$ denotes the entropy, (4.18a) is derived according to the definition of mutual information [3], and (4.18b) holds due to the independence between symbol modulation and subcarrier activation. Towards computing (4.14), we turn to calculate $E_{\mathbf{h}}[\mathrm{H}(\mathbf{x}_s, \mathbf{x}_c|\mathbf{y}, \mathbf{h})]$ in favor of (4.18b). First, let us express $\mathrm{H}(\mathbf{x}_s, \mathbf{x}_c|\mathbf{y}, \mathbf{h})$ according to its definition [3] as

$$\mathrm{H}(\mathbf{x}_s, \mathbf{x}_c|\mathbf{y}, \mathbf{h}) = -\sum_{j=1}^{\mathrm{C}(L,m)}\sum_{p^{(L-m)}}\int_{\mathbf{y}} \Pr\left(\mathbf{x}_c = \Omega_j, \mathbf{x}_s = \mathbf{s}\right) f\left(\mathbf{y}|\mathbf{x}_c = \Omega_j, \mathbf{x}_s = \mathbf{s}, \mathbf{h}\right) \times \log_2\left(\frac{\Pr\left(\mathbf{x}_c = \Omega_j, \mathbf{x}_s = \mathbf{s}\right) f\left(\mathbf{y}|\mathbf{x}_c = \Omega_j, \mathbf{x}_s = \mathbf{s}, \mathbf{h}\right)}{f(\mathbf{y}|\mathbf{h})}\right) d\mathbf{y}, \quad (4.19)$$

where $\mathbf{s} = [s_{p_1}, \ldots, s_{p_{L-m}}]^T \in \mathcal{X}^{L-m}$, $\Pr\left(\mathbf{x}_c = \Omega_j, \mathbf{x}_s = \mathbf{s}\right) = M^{m-L}/\mathrm{C}(L,m)$,

$$f\left(\mathbf{y}|\mathbf{x}_c = \Omega_j, \mathbf{x}_s = \mathbf{s}, \mathbf{h}\right) = \frac{1}{\left(\pi\sigma^2\right)^L}\prod_{i=1}^{m} e^{-\frac{\left|y_{\bar{\Omega}_j(i)}\right|^2}{\sigma^2}}\prod_{i=1}^{L-m} e^{-\frac{\left|y_{\Omega_j(i)} - h_{\Omega_j(i)}\sqrt{\frac{P_G}{L-m}} s_{p_i}\right|^2}{\sigma^2}}, \quad (4.20)$$

and

$$
\begin{aligned}
f\left(\mathbf{y}|\mathbf{h}\right) = & \frac{1}{\mathrm{C}\left(L,m\right)M^{L-m}\left(\pi\sigma^{2}\right)^{L}}\sum_{j'=1}^{\mathrm{C}(L,m)}\sum_{p'^{(L-m)}} \\
& \times\prod_{i'=1}^{L-m}e^{-\frac{\left|y_{\Omega_{j'}(i')}-h_{\Omega_{j'}(i')}\sqrt{\frac{P_G}{L-m}}s_{p'_{i'}}\right|^{2}}{\sigma^{2}}}\prod_{i'=1}^{m}e^{-\frac{\left|y_{\bar{\Omega}_{j'}(i')}\right|^{2}}{\sigma^{2}}}.
\end{aligned}
\tag{4.21}
$$

Then, changing the integral variables in (4.19) by

$$
v_i = \begin{cases} \left(y_i - h_i x_i\right)/\sigma, & i \in \Omega_j \\ y_i/\sigma, & i \in \bar{\Omega}_j \end{cases},
\tag{4.22}
$$

(4.19) can be reformulated as

$$
\begin{aligned}
\mathrm{H}\left(\mathbf{x}_s, \mathbf{x}_c|\mathbf{y}, \mathbf{h}\right) = & -L\log_2\left(\pi\right) + \frac{1}{M^{L-m}\mathrm{C}\left(L,m\right)}\sum_{j=1}^{\mathrm{C}(L,m)}\sum_{p^{(L-m)}} \\
& \times\int_{\mathbf{v}} f\left(\mathbf{v}\right)\log_2\left(\sum_{j'=1}^{\mathrm{C}(L,m)}\sum_{p'^{(L-m)}}\prod_{i=1}^{L}\frac{\xi\left(v_i,h_i,\gamma\right)}{2}\right)d\mathbf{v} \\
& + \frac{1}{M^{L-m}\mathrm{C}\left(L,m\right)}\sum_{j=1}^{\mathrm{C}(L,m)}\sum_{p^{(L-m)}}\int_{\mathbf{v}} f\left(\mathbf{v}\right)\log_2\left(\frac{1}{f\left(\mathbf{v}\right)}\right)d\mathbf{v},
\end{aligned}
\tag{4.23}
$$

where $\mathbf{v} = \left[v_1, \ldots, v_L\right]^T$ and $f\left(\mathbf{v}\right) = \pi^{-L}e^{-\|\mathbf{v}\|^2}$ from (4.22). Since extracting the coefficient $1/2$ from the first term at the right hand side of (4.23) yields $-L$ and the integral in the second term gives the differential entropy of $\mathbf{v}$, $\mathrm{h}\left(\mathbf{v}\right) = L\log_2\left(\pi e\right)$, we have

$$
\frac{1}{L}E_{\mathbf{h}}\left[\mathrm{H}\left(\mathbf{x}_s, \mathbf{x}_c|\mathbf{y}, \mathbf{h}\right)\right] = \log_2\left(e\right) - 1 + \lambda\left(\gamma\right).
\tag{4.24}
$$

Finally, substituting (4.24) into (4.18b) completes the proof. □

For the case that OFDM-IM has full subcarrier-correlations, i.e., $h_1 = \cdots = h_L$, the expectation over $\mathbf{h}$ in (4.15) is simplified to an expectation over a single random variable. However, it should be noted that the resulting achievable rate is larger than $\log_2(M)(L-m)/L$ at high SNR, which will be validated in the following Corollary 1.

Corollary 1 *When SNR approaches 0 and* $+\infty$*, the achievable rates of OFDM-IM tend to*

$$
\lim_{\gamma\to 0} R_{\mathrm{OFDM-IM}}\left(\gamma\right) = 0,
\tag{4.25}
$$

and

$$\lim_{\gamma\to+\infty} R_{\text{OFDM-IM}}(\gamma) = \frac{L-m}{L}\log_2(M) + \frac{\log_2(\mathrm{C}(L,m))}{L}, \tag{4.26}$$

respectively.

Proof First, consider $\gamma \to 0$. In this case, we have $\lim_{\gamma\to 0} \xi(v_i, h_i, \gamma) = 2e^{-|v_i|^2}$ from (4.17) and

$$\begin{aligned}\lim_{\gamma\to 0} \lambda(\gamma) =& \frac{1}{LM^{L-m}\mathrm{C}(L,m)} \sum_{j=1}^{\mathrm{C}(L,m)} \sum_{p^{(L-m)}} \\ &\times E_{\mathbf{v}}\left[\log_2\left(2^L e^{-\|\mathbf{v}\|^2}\mathrm{C}(L,m)\, M^{L-m}\right)\right] \\ =& \frac{L-m}{L}\log_2(M) + \frac{\log_2(\mathrm{C}(L,m))}{L} - \left(\log_2(e) - 1\right),\end{aligned} \tag{4.27}$$

from (4.16). Then, substituting (4.27) into (4.15) yields (4.25). Second, consider $\gamma \to +\infty$. Let $\mu\left(j, j', \mathbf{p}, \mathbf{p}'\right) = \left|j - j'\right| + \left\|\mathbf{p} - \mathbf{p}'\right\|$, where $\mathbf{p} = [p_1, \ldots, p_{L-m}]^T$ and $\mathbf{p}' = [p'_1, \ldots, p'_{L-m}]^T$. Then, (4.16) can be rewritten as

$$\begin{aligned}\lambda(\gamma) =& \frac{1}{LM^{L-m}\mathrm{C}(L,m)} \sum_{j=1}^{\mathrm{C}(L,m)} \sum_{p^{(L-m)}} \\ &\times E_{\mathbf{v},\mathbf{h}}\left[\log_2\left(2^L e^{-\|\mathbf{v}\|^2} + \underbrace{\sum_{j'=1}^{\mathrm{C}(L,m)} \sum_{p'^{(L-m)}}}_{\mu(j,j',\mathbf{p},\mathbf{p}')\neq 0} \prod_{i=1}^{L} \xi(v_i, h_i, \gamma)\right)\right].\end{aligned} \tag{4.28}$$

Since for $\mu\left(j, j', \mathbf{p}, \mathbf{p}'\right) \neq 0$ we always have at least an element, i, which belongs to $\left(\Omega_j \cap \Omega_{j'}\right) \cup \left(\Omega_j \cap \bar{\Omega}_{j'}\right) \cup \left(\bar{\Omega}_j \cap \Omega_{j'}\right)$, satisfying $\lim_{\gamma\to+\infty} \xi(v_i, h_i, \gamma) = 0$ from (4.17), the second term inside the logarithmic function in (4.28) approaches zero for $\gamma \to +\infty$. Therefore, it follows that $\lim_{\gamma\to+\infty} \lambda(\gamma) = 1 - \log_2(e)$, which, when substituted into (4.15), makes (4.26) valid. □

Corollary 1 is intuitive. When SNR is very small, the transmitted information is seriously contaminated by noise and thus no rate can be guaranteed by the channel. On the contrary, when SNR is very large, all transmitted information can be transferred through the channel with an arbitrarily low probability of error and thus the achievable rate is limited to the entropy at the channel input.

To the best of our knowledge, no closed-form result exists for (4.15). Consequently, we have to calculate it based on the Monte Carlo method, which, however, need not guarantee accuracy. To circumvent the above problems, we derive a closed-form lower bound in the sequel.

Proposition 2 *The achievable rate of OFDM-IM is lower bounded by*

$$R_{\text{OFDM-IM}}^{\text{L}}(\gamma) \triangleq \frac{L-m}{L}\log_2(M) + \frac{\log_2(\text{C}(L,m))}{L} - \log_2(e) + 1 - \frac{1}{L\text{C}(L,m)} \times \frac{1}{M^{L-m}} \sum_{j=1}^{\text{C}(L,m)} \sum_{p^{(L-m)}} \log_2\left(\sum_{j'=1}^{\text{C}(L,m)} \sum_{p'^{(L-m)}} \frac{1}{\left|\mathbf{I}_L + \Sigma\Lambda_{j,j'}\right|}\right), \quad (4.29)$$

where $\Lambda_{j,j'}$ is an $L \times L$ diagonal matrix whose i-th diagonal element is given by

$$\left[\Lambda_{j,j'}\right]_{i,i} = \begin{cases} \frac{\gamma L}{L-m}\frac{\left|s_{p_{\Omega_j^{-1}(i)}} - s_{p'_{\Omega_{j'}^{-1}(i)}}\right|^2}{2}, & i \in \Omega_j \cap \Omega_{j'} \\ \frac{\gamma L}{L-m}\frac{\left|s_{p_{\Omega_j^{-1}(i)}}\right|^2}{2}, & i \in \Omega_j \cap \bar{\Omega}_{j'} \\ \frac{\gamma L}{L-m}\frac{\left|s_{p'_{\Omega_{j'}^{-1}(i)}}\right|^2}{2}, & i \in \bar{\Omega}_j \cap \Omega_{j'} \\ 0, & i \in \bar{\Omega}_j \cap \bar{\Omega}_{j'} \end{cases}, \quad (4.30)$$

with $i = 1, \ldots, L$.

Proof The proof is somewhat similar to that in [4, Appendix B]. Thus, we only provide some intermediate steps toward the result. Since $\log_2(\cdot)$ is a concave function, applying Jensen's inequality to (4.16) gives

$$\lambda(\gamma) \leq \frac{1}{LM^{L-m}\text{C}(L,m)} \sum_{j=1}^{\text{C}(L,m)} \sum_{p^{(L-m)}} \times \log_2\left(\sum_{j'=1}^{\text{C}(L,m)} \sum_{p'^{(L-m)}} E_{\mathbf{h}}\left[\prod_{i=1}^{L} E_{v_i}\left[\xi(v_i, h_i, \gamma)\right]\right]\right), \quad (4.31)$$

where we have resorted to the property that $\mathbf{v}$ is independent of $\mathbf{h}$. Thanks to the integral result in [5, Eq. (3.323.2)], the inner expectation in (4.31) can be solved as

$$E_{v_i}\left[\xi(v_i, h_i, \gamma)\right] = \begin{cases} e^{-\frac{\gamma L}{L-m}\frac{|h_i|^2\left|s_{p_{\Omega_j^{-1}(i)}} - s_{p'_{\Omega_{j'}^{-1}(i)}}\right|^2}{2}}, & i \in \Omega_j \cap \Omega_{j'} \\ e^{-\frac{\gamma L}{L-m}\frac{|h_i|^2\left|s_{p_{\Omega_j^{-1}(i)}}\right|^2}{2}}, & i \in \Omega_j \cap \bar{\Omega}_{j'} \\ e^{-\frac{\gamma L}{L-m}\frac{|h_i|^2\left|s_{p'_{\Omega_{j'}^{-1}(i)}}\right|^2}{2}}, & i \in \bar{\Omega}_j \cap \Omega_{j'} \\ 1, & i \in \bar{\Omega}_j \cap \bar{\Omega}_{j'} \end{cases}. \quad (4.32)$$

Then, introducing the matrix $\Lambda_{j,j'}$, we can express $\prod_{i=1}^{L} E_{v_i}\left[\xi\left(v_i, h_i, \gamma\right)\right]$ according to (4.32) as $e^{-\mathbf{h}^H \Lambda_{j,j'}\mathbf{h}}$. Finally, the expectation over $\mathbf{h}$ in (4.31) can be solved in closed form as

$$E_{\mathbf{h}}\left[\prod_{i=1}^{L} E_{v_i}\left[\xi\left(v_i, h_i, \gamma\right)\right]\right] = \frac{1}{\pi^L|\Sigma|}\int_{\mathbf{h}} e^{-\mathbf{h}^H\left(\Sigma^{-1}+\Lambda_{j,j'}\right)\mathbf{h}} d\mathbf{h}$$
$$= \left|\mathbf{I}_L + \Sigma\Lambda_{j,j'}\right|^{-1}, \tag{4.33}$$

which, when substituted into (4.31) and then into (4.15), proves (4.29). □

Corollary 2 *When all subcarriers within a subcarrier group are faded completely correlatively, the proposed lower bound becomes*

$$R^{\mathrm{L}}_{\mathrm{OFDM-IM-FC}}(\gamma) \triangleq \frac{L-m}{L}\log_2(M) + \frac{\log_2\left(\mathrm{C}(L,m)\right)}{L} - \left(\log_2(e) - 1\right)$$
$$- \frac{1}{L\mathrm{C}(L,m)M^{L-m}} \sum_{j=1}^{\mathrm{C}(L,m)} \sum_{p^{(L-m)}}$$
$$\times \log_2\left(\sum_{j'=1}^{\mathrm{C}(L,m)} \sum_{p'^{(L-m)}} \frac{1}{1 + Tr\left\{\Lambda_{j,j'}\right\}}\right). \tag{4.34}$$

Proof In the case of $h_1 = \cdots = h_L$, Σ becomes an all-ones matrix and $\left|\mathbf{I}_L + \Sigma\Lambda_{j,j'}\right| = 1 + Tr\left\{\Lambda_{j,j'}\right\}$, which, when substituted into (4.29), proves (4.34). □

Corollary 3 *When SNR approaches 0 and $+\infty$, the proposed lower bounds tend to*

$$\lim_{\gamma\to 0} R^{\mathrm{L}}_{\mathrm{OFDM-IM}}(\gamma) = -\log_2(e) + 1, \tag{4.35}$$

and

$$\lim_{\gamma\to +\infty} R^{\mathrm{L}}_{\mathrm{OFDM-IM}}(\gamma) = \frac{L-m}{L}\log_2(M) + \frac{\log_2\left(\mathrm{C}(L,m)\right)}{L} - \left(\log_2(e) - 1\right), \tag{4.36}$$

respectively.

Proof For $\gamma \to 0$, we have $\Lambda_{j,j'} \to \mathbf{0}_L$. Therefore, (4.35) holds from (4.29). On the other hand, for $\gamma \to +\infty$, we have

$$\lim_{\gamma\to+\infty}\sum_{j'=1}^{C(L,m)}\sum_{p'^{(L-m)}}\frac{1}{\left|\mathbf{I}_L+\Sigma\Lambda_{j,j'}\right|}$$
$$=1+\lim_{\gamma\to+\infty}\underbrace{\sum_{j'=1}^{C(L,m)}\sum_{p'^{(L-m)}}}_{\mu(j,j',\mathbf{p},\mathbf{p}')\neq 0}\frac{1}{\left|\mathbf{I}_L+\Sigma\Lambda_{j,j'}\right|}=1. \tag{4.37}$$

Therefore, (4.36) holds from (4.29). □

From Corollarys 1 and 3, a constant gap, i.e., $-\log_2(e)+1$, between $R_{\text{OFDM-IM}}(\gamma)$ and $R^{\text{L}}_{\text{OFDM-IM}}(\gamma)$ will be expected in both low and high SNR regions. However, this does not apply to the remaining SNR region. Despite this, $R^{\text{L}}_{\text{OFDM-IM}}(\gamma)$ tracks very well the trend of $R_{\text{OFDM-IM}}(\gamma)$ with the increase of SNR and can be used to estimate some parameters of practical interest with satisfactory accuracy.

From (4.29), it is clear that the achievable rate of OFDM-IM not only depends on the channel correlation, which is controlled by Σ, but also on the modulation type, which determines $\Lambda_{j,j'}$. An example for the former issue has been presented in Corollary 2. In the following, both issues will be discussed in more detail.

When only subcarrier activation is applied in OFDM-IM, which is called OFDM with subcarrier shift keying (OFDM-SSK), the resulting achievable rate can be expressed as

$$\begin{aligned}R_{\text{OFDM-SSK}}(\gamma)=&\frac{\log_2\left(\mathrm{C}(L,m)\right)}{L}-\left(\log_2(e)-1\right)-\frac{1}{L\mathrm{C}(L,m)}\sum_{j=1}^{\mathrm{C}(L,m)}\\&\times E_{\mathbf{v},\mathbf{h}}\left[\log_2\left(\sum_{j'=1}^{\mathrm{C}(L,m)}\prod_{i\in(\Omega_j\cap\Omega_{j'})\cup(\bar{\Omega}_j\cap\bar{\Omega}_{j'})}2e^{-|v_i|^2}\right.\right.\\&\left.\left.\times\prod_{i\in\Omega_j\cap\bar{\Omega}_{j'}}2e^{-\left|v_i+h_i\sqrt{\frac{\gamma L}{L-m}}\right|^2}\prod_{i\in\bar{\Omega}_j\cap\Omega_{j'}}2e^{-\left|v_i-h_i\sqrt{\frac{\gamma L}{L-m}}\right|^2}\right)\right],\end{aligned} \tag{4.38}$$

which follows from (4.15) by setting $M=1$. Also, a closed-form lower bound exists by letting $M=1$ in (4.29):

$$\begin{aligned}R^{\text{L}}_{\text{OFDM-SSK}}(\gamma)=&\frac{\log_2\left(\mathrm{C}(L,m)\right)}{L}-\left(\log_2(e)-1\right)-\frac{1}{L\mathrm{C}(L,m)}\\&\times\sum_{j=1}^{\mathrm{C}(L,m)}\log_2\left(\sum_{j'=1}^{\mathrm{C}(L,m)}\frac{1}{\left|\mathbf{I}_L+\Sigma\Gamma_{j,j'}\right|}\right),\end{aligned} \tag{4.39}$$

where $\Gamma_{j,j'}$ is an $L\times L$ diagonal matrix whose i-th diagonal element is

$$\left[\Gamma_{j,j'}\right]_{i,i} = \begin{cases} 0, & i \in \left(\Omega_j \cap \Omega_{j'}\right) \cup \left(\bar{\Omega}_j \cap \bar{\Omega}_{j'}\right) \\ \frac{\gamma L}{2(L-m)}, & i \in \left(\bar{\Omega}_j \cap \Omega_{j'}\right) \cup \left(\Omega_j \cap \bar{\Omega}_{j'}\right) \end{cases}, \tag{4.40}$$

with $i = 1, \ldots, L$.

Recalling that on-off keying (OOK) transmits information by the intensity of the signal [6], we can readily determine that the OFDM-SSK is in fact an application of OOK modulation to OFDM subcarriers. Similarly, we can also apply OOK modulation to each subcarrier independently, thus generating a new scheme called OOK-OFDM. In Sect. 4.1.3, we will compare the performance of the two aforementioned schemes.

Infinite Constellation Input

Let us consider $\mathbf{x}_s$ to be an infinite signal set, i.e., M is infinite. However, we restrict the signal set to be Gaussian, since it enables OFDM-IM to reach the supremum of all achievable rates [3, 7].

To begin with the derivation for the ergodic capacity of OFDM-IM, we first look into the conditional capacity of OFDM-IM on the channel, which can be expressed according to the chain rule for mutual information [3] as

$$C_{\text{OFDM-IM}}(\gamma) = \frac{1}{L}\mathrm{I}\left(\mathbf{x}_s, \mathbf{x}_c; \mathbf{y}|\mathbf{h}\right) = \frac{1}{L}\left[\mathrm{I}\left(\mathbf{x}_s; \mathbf{y}|\mathbf{x}_c, \mathbf{h}\right) + \mathrm{I}\left(\mathbf{x}_c; \mathbf{y}|\mathbf{h}\right)\right]. \tag{4.41}$$

In (4.41), the first term at the right hand side is given by

$$\begin{aligned} \frac{1}{L}\mathrm{I}\left(\mathbf{x}_s; \mathbf{y}|\mathbf{x}_c, \mathbf{h}\right) &= \frac{1}{L}\sum_{j=1}^{\mathrm{C}(L,m)} \Pr\left(\mathbf{x}_c = \Omega_j\right) \mathrm{I}\left(\mathbf{x}_s; \mathbf{y}|\mathbf{x}_c = \Omega_j, \mathbf{h}\right) \\ &= \frac{1}{L\mathrm{C}(L,m)} \sum_{j=1}^{\mathrm{C}(L,m)} \sum_{i\in\Omega_j} \log_2\left(1 + \frac{\gamma |h_i|^2 L}{L-m}\right) \\ &= \frac{L-m}{L}\left[\frac{1}{L}\sum_{i=1}^{L} \log_2\left(1 + \frac{\gamma |h_i|^2 L}{L-m}\right)\right], \end{aligned} \tag{4.42}$$

where the last equality follows from the fact that each subcarrier is activated with probability $(L-m)/L$. It is clear from (4.42) that the degrees of freedom associated with OFDM-IM reduce to $L-m$ as opposed to L associated with classical OFDM, but OFDM-IM contributes a power gain of $L/(L-m) > 1$. From (4.41), we see that in addition to this power gain, OFDM-IM also introduces a portion of information rate from subcarrier activation, i.e., the second term at the right hand side:

$$\frac{\mathrm{I}\left(\mathbf{x}_c; \mathbf{y}|\mathbf{h}\right)}{L} = \frac{1}{L\mathrm{C}(L,m)} \sum_{j=1}^{\mathrm{C}(L,m)} \int_{\mathbf{y}} f\left(\mathbf{y}|\mathbf{x}_c = \Omega_j, \mathbf{h}\right) \log_2\left(\frac{f\left(\mathbf{y}|\mathbf{x}_c = \Omega_j, \mathbf{h}\right)}{f\left(\mathbf{y}|\mathbf{h}\right)}\right) d\mathbf{y}, \tag{4.43}$$

where

$$f\left(\mathbf{y}|\mathbf{x}_c=\Omega_j,\mathbf{h}\right)=\frac{e^{-\frac{|y_i|^2}{\sigma^2\left[1+\gamma|h_i|^2L/(L-m)\right]}}}{\left(\pi\sigma^2\right)^L}\prod_{i\in\Omega_j}\frac{L-m}{L-m+\gamma|h_i|^2L}\prod_{i\in\bar{\Omega}_j}e^{-\frac{|y_i|^2}{\sigma^2}}, \quad (4.44)$$

and

$$\begin{aligned}f\left(\mathbf{y}|\mathbf{h}\right)=&\sum_{j=1}^{\mathrm{C}(L,m)}\Pr\left(\mathbf{x}_c=\Omega_j\right)f\left(\mathbf{y}|\mathbf{x}_c=\Omega_j,\mathbf{h}\right)\\=&\frac{1}{\mathrm{C}\left(L,m\right)\left(\pi\sigma^2\right)^L}\sum_{j=1}^{\mathrm{C}(L,m)}\prod_{i\in\bar{\Omega}_j}e^{-\frac{|y_i|^2}{\sigma^2}}\\&\times\prod_{i\in\Omega_j}\frac{1}{1+\gamma|h_i|^2L/\left(L-m\right)}e^{-\frac{|y_i|^2}{\sigma^2\left[1+\gamma|h_i|^2L/(L-m)\right]}}.\end{aligned} \quad (4.45)$$

According to the above analysis, one can expect that the instantaneous capacity of OFDM-IM will be smaller than that of the classical OFDM at high SNR due to the reduction in degrees of freedom. However, as OFDM-IM benefits from a power gain and additional information rate transferred via subcarrier activation, it is not clear whether OFDM-IM can outperform classical OFDM or not at low-to-medium SNR. The answer is given in the following proposition.

Proposition 3 *The instantaneous capacity of OFDM-IM is smaller than that of classical OFDM over the entire SNR region:*

$$C_{\mathrm{OFDM-IM}}\left(\gamma\right)<C_{\mathrm{OFDM}}\left(\gamma\right),\ \forall m>0, \quad (4.46)$$

where

$$C_{\mathrm{OFDM}}\left(\gamma\right)=\frac{1}{L}\sum_{i=1}^{L}\log_2\left(1+\gamma|h_i|^2\right), \quad (4.47)$$

denotes the instantaneous capacity of classical OFDM [1, 2].

Proof Under Gaussian input, the PDF of the received signal on the i-th subcarrier conditioned on the channel is given by

$$f\left(y_i|h_i\right)=\frac{1}{\pi\sigma^2}\left[\frac{\left(L-m\right)^2}{L}\frac{1}{L-m+\gamma|h_i|^2L}e^{-\frac{(L-m)|y_i|^2}{\sigma^2\left(L-m+\gamma|h_i|^2L\right)}}+\frac{m}{L}e^{-\frac{|y_i|^2}{\sigma^2}}\right], \quad (4.48)$$

where $i = 1, \ldots, L$. In light of (4.48), it can be readily proved that the mean of y_i conditioned on h_i is zero and its variance yields

$$E_{y_i|h_i}\left[|y_i|^2\right] = \frac{L-m}{L}\sigma^2\left(1+\frac{\gamma|h_i|^2 L}{L-m}\right) + \frac{m}{L}\sigma^2 = \frac{|h_i|^2 P_G}{L} + \sigma^2. \tag{4.49}$$

Since the differential entropy of a zero-mean non-Gaussian RV is always smaller than that of a zero-mean Gaussian RV of the same variance [3], it follows that

$$\mathrm{h}\left(y_i|h_i\right) \leq \log_2\left(\pi e \sigma^2\left(\gamma|h_i|^2+1\right)\right), \tag{4.50}$$

with equality iff $m = 0$ which corresponds to the classical OFDM case. Then, according to [3, Theorem 9.6.2], i.e.,

$$\mathrm{h}\left(\mathbf{y}|\mathbf{h}\right) \leq \mathrm{h}\left(y_1|h_1\right) + \cdots + \mathrm{h}\left(y_L|h_L\right), \tag{4.51}$$

with equality iff the received signals at all subcarriers conditioned on the channel are independent, and $\mathrm{I}\left(\mathbf{x}_s, \mathbf{x}_c; \mathbf{y}|\mathbf{h}\right) = \mathrm{h}\left(\mathbf{y}|\mathbf{h}\right) - L\log_2\left(\pi e \sigma^2\right)$, we arrive at (4.46), completing the proof. □

In OFDM-IM, the power of the received signal on each subcarrier remains the same as that of classical OFDM. The distribution of the received signal, however, is not Gaussian. This differs from classical OFDM, thus leading to a smaller instantaneous capacity.

Having the conditional capacity of OFDM-IM in (4.41), we can then calculate the ergodic capacity of OFDM-IM, denoted by $\bar{C}_{\mathrm{OFDM-IM}}(\gamma)$, by averaging the channel realizations. Numerical results will be presented in Sect. 4.1.3. From Proposition 3, it is expected that the ergodic capacity of OFDM-IM will always be smaller than that of classical OFDM. This can be explained alternatively by the fact that the parallel channels resulting from OFDM-IM are rank-reduced, while the ergodic capacity of a vector Gaussian channel is achieved if and only if the channel has full rank.

4.1.3 Performance Evaluation

In this subsection, we present numerical results to make comparisons between classical OFDM and OFDM-IM. In the numerical calculations, the total number of subcarriers is chosen as $N = 64$ and the channel is assumed to have an exponentially decaying power delay profile with $\delta_d^2 = e^{-(d-1)}/\sum_{u=1}^{D} e^{-(u-1)}$ and $D = 10$ taps, where $d = 1, \ldots, D$. The achievable rate is chosen as the performance metric. Since as will be validated in Sect. 4.2.1, the ultimate performance of OFDM-IM is available by adopting interleaved grouping in practice, in the following we will simply let the subcarriers within each subcarrier group experience independent fading in order to examine the ultimate performance of OFDM-IM and avoid channel particularity.

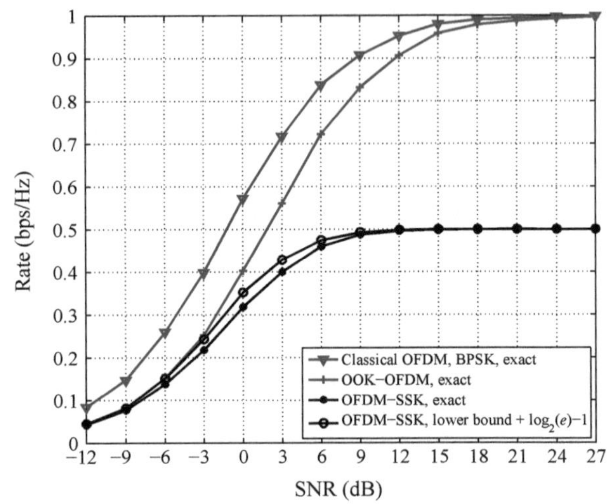

Fig. 4.2 Performance comparison between classical OFDM with BPSK, OOK-OFDM, and OFDM-SSK

Performance of OFDM-SSK

With the aid of (4.38) and (4.39), we compute the achievable rates of OFDM-SSK, classical OFDM with BPSK, and OOK-OFDM, in Fig. 4.2. To examine the constant-shifted method mentioned in Sect. 4.1.2, we add $\log_2(e) - 1$ to the lower bound on the achievable rate of OFDM-SSK. As seen from the figure, this constant-shifted lower bound overestimates the exact value in the moderate SNR region, though it achieves a perfect approximation at both low and high SNRs. Therefore, this method is infeasible in practice. Regarding the performance of these three schemes, we can conclude from Fig. 4.2 that classical OFDM with BPSK outperforms the other two schemes while OFDM-SSK performs worse in general though it exhibits similar performance to OOK-OFDM in the very low SNR region. The positive result associated with classical OFDM with BPSK can be accounted for its 3 dB larger minimum Euclidean distance than OOK-OFDM and its double entropy at the channel input over OFDM-SSK. The behavior of OFDM-SSK in the very low SNR region can be attributed to the coding gain from index modulation.

Performance of Different Subcarrier Activation Strategies

We present an example with $L = 8$ and 8-PSK in Fig. 4.3, where m varies from 1 to 5. As can be seen, besides $m = 1$, in which case OFDM-IM and classical OFDM have the same entropy at the channel input, OFDM-IM with $m = 2$ and $m = 3$ also have the potential to outperform classical OFDM despite smaller input entropy. For $m = 4$ and 5, the input entropies are too small to exploit the channel, thus leading to a smaller achievable rate than classical OFDM. Therefore, an intriguing question may arise: what is the optimal m in maximizing the achievable rate for a given SNR. For the above example, answers are given in Fig. 4.4, where SNRs equal to 0, 6, and 12 dB are considered, and the corresponding lower bounds are also plotted. It is shown that there exists an optimal non-zero m (marked with a circle) which may vary

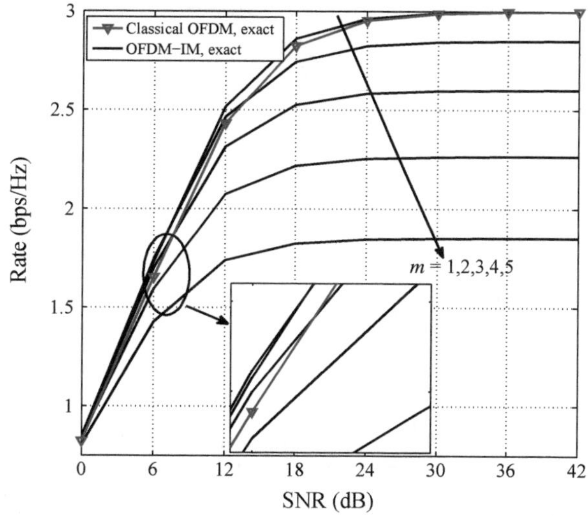

Fig. 4.3 Comparison between classical OFDM and OFDM-IM with $L = 8$ and 8-PSK

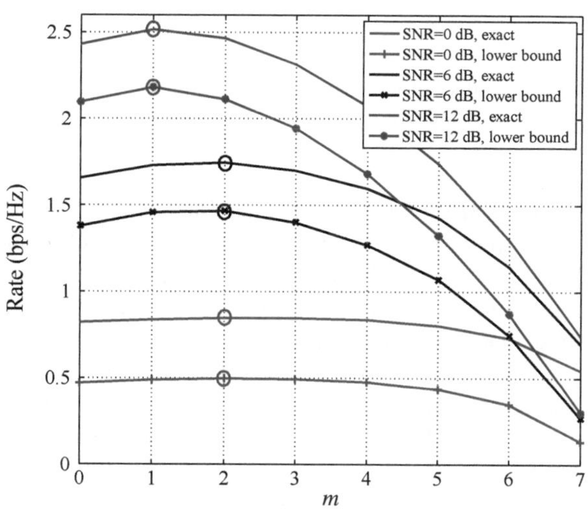

Fig. 4.4 Effect of m on the achievable rate in the case of $L = 8$, 8-PSK, and SNRs equal to 0, 6, and 12 dB, respectively

for different SNR values. However, we expect in general that when SNR becomes very high the optimal m is more likely to be 1 since the entropy at the channel input is the largest. On the other hand, it is clear from Fig. 4.4 that the proposed lower bound can also point out the same solutions for its potential to trace the exact values as the increase of SNR well (see Fig. 4.2 for further evidences). Therefore, for computational tractability, the use of the proposed lower bound is more advisable in the determination of the optimal m. It is worth noting that under a finite-alphabet input, the achievable rates of all schemes will always saturate at their corresponding uncoded transmitted information rates at very high SNR.

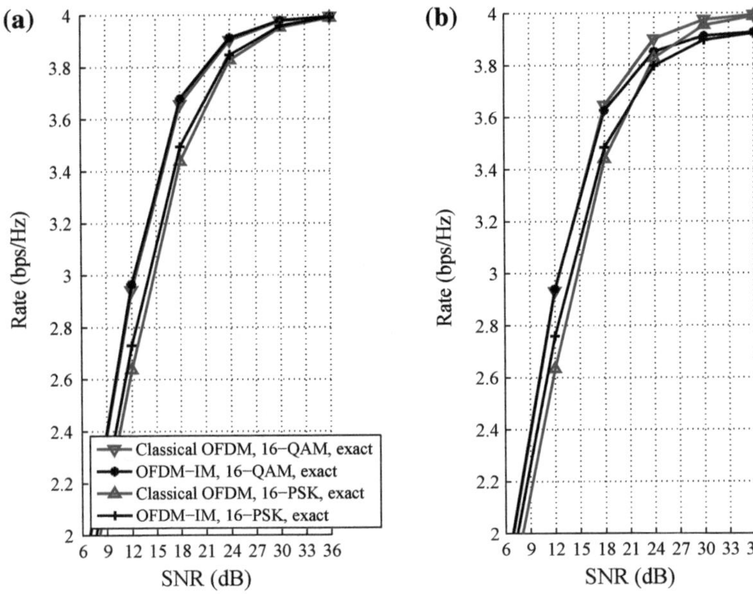

Fig. 4.5 Effect of modulation types on the achievable rates of classical OFDM and OFDM-IM with $L = 16$ and **a** $m = 1$; **b** $m = 2$

Which is More Favorable For OFDM-IM: PSK or QAM?
An example for the comparison between 16-PSK and 16-QAM inputs is shown in Fig. 4.5, where the simulation parameters are $L = 16$ and (a) $m = 1$; (b) $m = 2$. For figure clarity, the lower bounds are omitted. Different from the conclusion drawn in spatial modulation that PSK may be better than QAM [8], from the figure we see that QAM is always more favorable than PSK for OFDM-IM in the sense of higher achievable rate. On the other hand, we see that OFDM-IM with $m = 1$ and $m = 2$ shows an SNR gain of nearly 1 dB over classical OFDM for a large SNR range when both employ 16-PSK modulation whereas very little gain is found when both employ 16-QAM modulation. Reasons can be found by comparing the minimum Euclidean distances of the overall constellations of classical OFDM and OFDM-IM with 16-PSK and 16-QAM. The overall constellations of OFDM-IM are different from those of classical OFDM. Particularly, in OFDM-IM, one more constellation point is located at the origin, which is introduced from the inactive state, and meanwhile a power gain of $L/(L-m)$ is shared by all conventional 16-PSK and 16-QAM constellation points. Therefore, the minimum Euclidean distance of the overall constellation of OFDM-IM with 16-PSK will still be determined by the two adjacent 16-PSK constellation points of enhanced power while that with 16-QAM will be determined by the origin point and its nearest 16-QAM constellation point of enhanced power. It should be noted that since in OFDM-IM the origin point appears with probability m/L while each 16-PSK/QAM constellation point appears with probability $(L-m)/ML$, the comparison through measuring the Euclidean

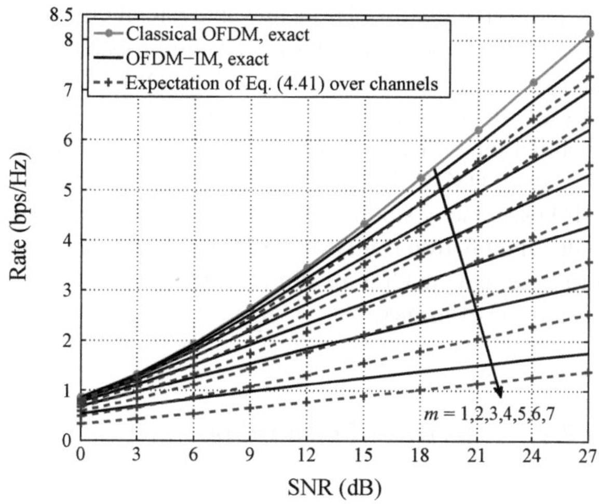

Fig. 4.6 Comparison between classical OFDM and OFDM-IM with $L = 8$ in terms of ergodic capacity

distance directly need not guarantee absolute accuracy. However, this can be adopted as a general principle to foresee the comparison result when a larger alphabet, such as 64-PSK/QAM is considered.

Effects of Gaussian Input

We compare the performance of OFDM-IM with classical OFDM under Gaussian input.

With $L = 8$ and m spanning from 1 to 7, the comparison in terms of ergodic capacity is illustrated in Fig. 4.6. The expectation of (4.42) over channels, which accounts for the portion of ergodic capacity contributed by symbol modulation, is also depicted. As seen from the figure, the contribution of subcarrier activation to the ergodic capacity, which equals the gap between the black and blue dashed curves, cannot be neglected. Take $m = 4$ for example. At SNR equal to 27 dB, the ergodic capacity of OFDM-IM is about 5.3 bps/Hz and the portion contributed by symbol modulation is about 4.6 bps/Hz. Therefore, the other portion contributed by subcarrier activation is about $5.3 - 4.6 = 0.7$ bps/Hz, resulting in the ratio between the portions of the ergodic capacity contributed by subcarrier activation and symbol modulation as high as $0.7/4.6 = 15.2\,\%$. On the other hand, it is clear that the figure substantiates our conclusion presented in Sect. 4.1.2 that OFDM-IM performs worse than classical OFDM under Gaussian input.

4.2 Design Guidelines for Performance Improvement

In this section, we propose four novel methods suited to OFDM-IM systems to improve the diversity gain, coding gain, and spectral efficiency, respectively. The first one is called interleaved grouping, which groups subcarriers in an interleaved manner for index modulation. The achievable rate of OFDM-IM with interleaved grouping is analyzed. The second one is called equiprobable subcarrier activation (ESA), which allows equiprobable activation of all OFDM-IM subcarriers. The ABEP performance of OFDM-IM systems with ESA is analyzed. The third one is called hybrid index modulation, which introduces an additional bit to select the classical OFDM mode or the OFDM-IM mode. The achievable rate of the hybrid OFDM-IM (OFDM-HIM) scheme is studied. The fourth one is called OFDM with in-phase/quadrature index modulation (OFDM-I/Q-IM), which extends index modulation to include the in-phase and quadrature dimensions. The ABEP performance of OFDM-I/Q-IM systems is examined.

4.2.1 Interleaved Grouping

In this subsection, we propose a new subcarrier grouping approach for OFDM-IM to improve the performance of OFDM-IM, whose idea is motivated by the derived lower bound on the achievable rate of OFDM-IM in Sect. 4.1.2, which reveals that the correlation among channel coefficients on the subcarriers within a subcarrier group plays a key role.

Unlike the previous OFDM-IM scheme, in which the subcarrier grouping is performed in a localized manner [9], we suggest grouping the subcarriers in an interleaved manner, as shown in Fig. 4.7. After interleaved grouping, the set of subcarrier indices for subcarrier group g will be given by

$$\begin{aligned}\Psi^g &= \left\{\beta_1^g, \ldots, \beta_L^g\right\} \\ &= \{g, \ldots, g + (L-1)\,G\}, \end{aligned} \tag{4.52}$$

with $\beta_l^g = g + (l-1)\,G$, where $g = 1, \ldots, G$ and $l = 1, \ldots, L$. Note that when interleaved grouping is employed at the transmitter, it should be followed at the receiver. It is evident that applying either subcarrier grouping rule, i.e., localized grouping or interleaved grouping, leads to the same overall system complexity. Also note that from (4.52), it is clear that interleaved grouping is a kind of regular subcarrier grouping approaches since the (a, b)-th entry of the covariance matrix of $\mathbf{h}^g$ is $[\Sigma_{\mathrm{IG}}]_{a,b} = \sum_{d=1}^{D} \delta_d^2 e^{-2\sqrt{-1}\pi(a-b)(d-1)/L}$ with $a, b = 1, \ldots, L$, which does not depend on g. Therefore, the analysis conducted in Sect. 4.1.2 also applies to interleaved grouping. The reason for adopting interleaved grouping is clarified in the following proposition.

Proposition 4 *OFDM-IM with interleaved grouping achieves the highest information rate in both low and high SNR regions when the spacing of any two subcarriers within a subcarrier group is no smaller than the coherence bandwidth of the channel.*

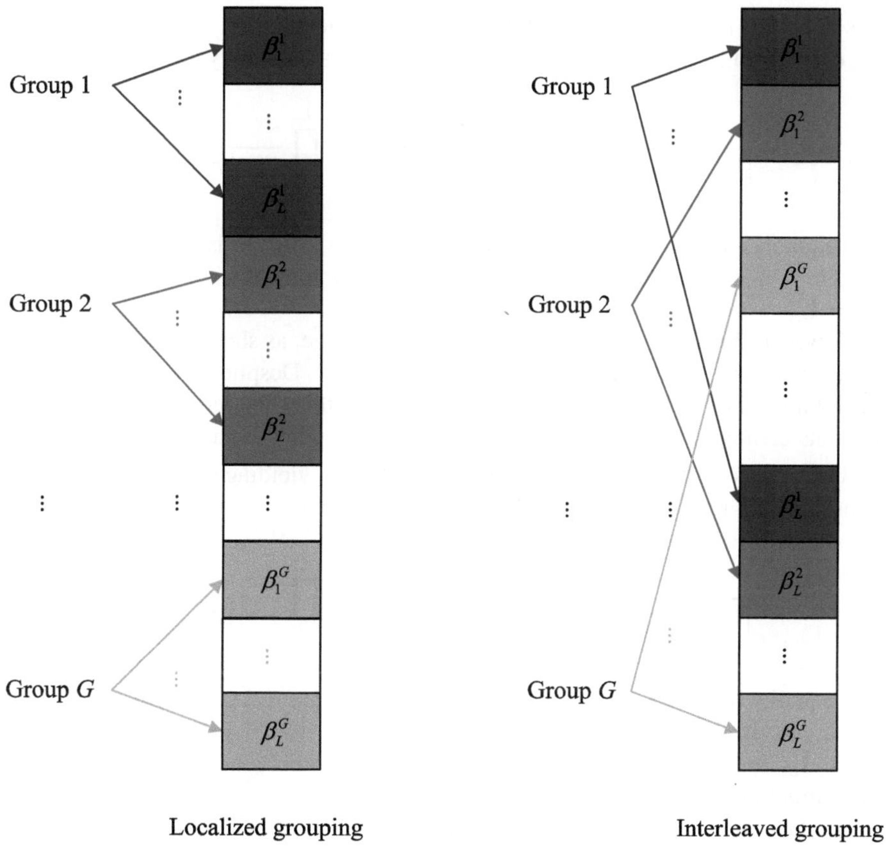

Fig. 4.7 Localized grouping versus interleaved grouping

Proof When SNR is either low or high, it follows from [10] that

$$R_{\mathrm{OFDM-IM}}(\gamma) \approx \frac{L-m}{L}\log_2(M) + \frac{\log_2(\mathrm{C}(L,m))}{L} - \frac{1}{L\mathrm{C}(L,m)M^{L-m}} \times \sum_{j=1}^{\mathrm{C}(L,m)} \sum_{p^{(L-m)}} \log_2\left(\sum_{j'=1}^{\mathrm{C}(L,m)} \sum_{p'^{(L-m)}} \frac{1}{\left|\mathbf{I}_L + \Sigma\Lambda_{j,j'}\right|}\right), \tag{4.53}$$

where $\Lambda_{j,j'}$ is defined in (4.30). Note that it is straightforward to have (4.53) according to Corollarys 1 and 3. From (4.53), the optimal Σ^*, which maximizes the achievable rate of OFDM-IM, lies in the following:

$$\Sigma^* = \arg\min_{\forall\Sigma} \sum_{j=1}^{\mathrm{C}(L,m)} \sum_{p^{(L-m)}} \log_2\left(\sum_{j'=1}^{\mathrm{C}(L,m)} \sum_{p'^{(L-m)}} \frac{1}{\left|\mathbf{I}_L + \Sigma\Lambda_{j,j'}\right|}\right). \tag{4.54}$$

To solve (4.54), let us assume that the diagonal elements of $\Lambda_{j,j'}$ are all positive. In this case, from [3, Theorem 16.8.2] (i.e., Hadamard's inequality), we have

$$\frac{1}{\left|\mathbf{I}_L + \Sigma\Lambda_{j,j'}\right|} = \frac{1}{\left|\Lambda_{j,j'}^{-1} + \Sigma\right| \left|\Lambda_{j,j'}\right|} \geq \prod_{i=1}^{L} \frac{1}{1 + \left[\Lambda_{j,j'}\right]_{i,i}}, \tag{4.55}$$

with equality iff $\Sigma = \mathbf{I}_L$, which represents the case in which $\mathbf{h}^g$ is a fully independent random vector. Since the logarithmic function is monotonically increasing, we obtain $\Sigma^* = \mathbf{I}_L$.

However, the above assumption does not hold since, as shown in (4.30), there is at least one zero among the diagonal elements of $\Lambda_{j,j'}$. Despite this, we will see that the solution, i.e., $\Sigma^* = \mathbf{I}_L$, is still valid. To show this, first assume that only one diagonal element of $\Lambda_{j,j'}$, e.g., $\left[\Lambda_{j,j'}\right]_{k,k}$, equals zero. In this case, we can eliminate the k-th row as well as the k-th column of Σ and $\Lambda_{j,j'}$, yielding updated $\widetilde{\Sigma}$ and $\widetilde{\Lambda}_{j,j'}$. Consequently, it follows from [3, Theorem 16.8.2] that

$$\frac{1}{\left|\mathbf{I}_L + \Sigma\Lambda_{j,j'}\right|} = \frac{1}{\left|\widetilde{\Lambda}_{j,j'}^{-1} + \widetilde{\Sigma}\right| \left|\widetilde{\Lambda}_{j,j'}\right|} \geq \prod_{i=1}^{L} \frac{1}{1 + \left[\Lambda_{j,j'}\right]_{i,i}}, \tag{4.56}$$

with equality iff $\widetilde{\Sigma} = \mathbf{I}_{L-1}$. Since for different $j,\ j' \in \{1, \ldots, \mathrm{C}(L, m)\}$, the location of the zero in the diagonal elements of $\Lambda_{j,j'}$ spans from 1 to L, we achieve that $\Sigma^* = \mathbf{I}_L$. In analogy with the above analysis, the same conclusion can be drawn when more diagonal elements of $\Lambda_{j,j'}$ equal zero.

In order to achieve the maximum rate of OFDM-IM, the subcarriers within a subcarrier group should be spaced apart in frequency as far as possible to benefit from the independent fading. It is easy to see that interleaved grouping meets this requirement most perfectly, thus completing the proof. □

In practical systems, the condition in Proposition 4 can be satisfied. Take LTE systems for example. In a macrocell, the coherence bandwidth is on the order of 1 MHz [11]. Therefore, for a system bandwidth of 20 MHz, the subcarriers within each subcarrier group after interleaved grouping will experience independent fading provided that $L < 20$. Since as disclosed in Sect. 4.1.3, the superiority of OFDM-IM over classical OFDM narrows as the alphabet size of the signal input grows, $L < 20$ is advantageous for OFDM-IM. Since $\bar{C}_{\text{OFDM-IM}}(\gamma)$ is dominated by $E_{\mathbf{h}}[\mathrm{I}(\mathbf{x}_s; \mathbf{y}|\mathbf{x}_c, \mathbf{h})]$ with $\mathbf{x}_s$ being a Gaussian signal set (especially as SNR gets higher), which is independent of the channel correlation as (4.42) implies, the gap between the resulting ergodic capacity of localized grouping and interleaved grouping will be very small. From (4.29) and the proof for Proposition 4, we see that the lower bound on the achievable rate of OFDM-IM achieves its maximum when $\Sigma = \mathbf{I}_L$, i.e., subcarriers within a subcarrier group are faded independently. Therefore, we expect that the interleaved

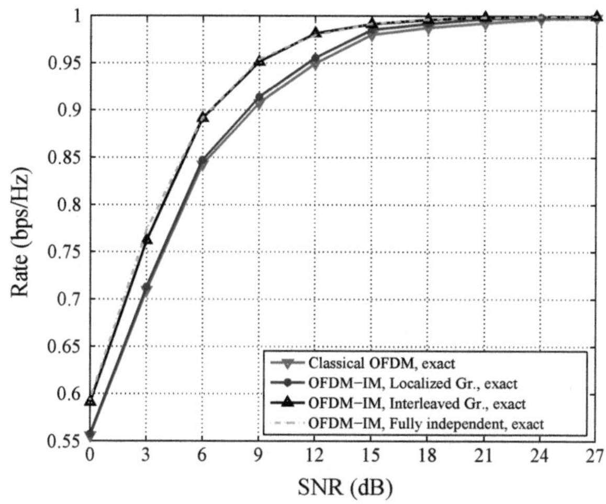

Fig. 4.8 Comparison between classical OFDM and OFDM-IM with localized grouping and interleaved grouping in terms of achievable rate with parameters chosen as $L = 2$, $m = 1$, and BPSK

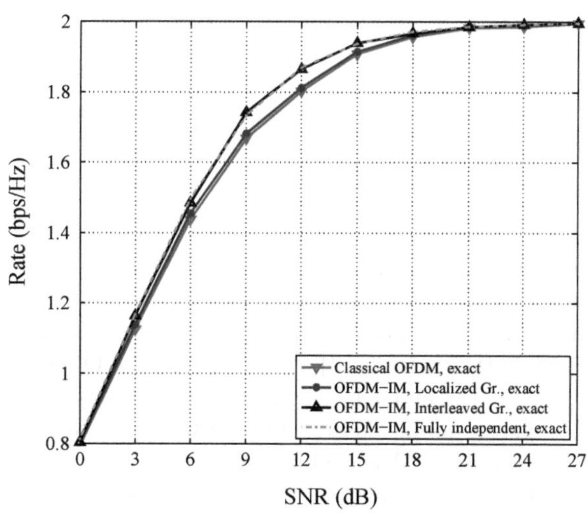

Fig. 4.9 Comparison between classical OFDM and OFDM-IM with localized grouping and interleaved grouping in terms of achievable rate with parameters chosen as $L = 4$, $m = 1$, and QPSK

grouping will also perform better than the localized grouping in the moderate SNR region for PSK/QAM-constrained input.

Here, we give an example to show how the interleaved grouping behaves with respect to the localized grouping. In this example, the total number of subcarriers is chosen as $N = 64$ and the channel is assumed to have an exponentially decaying power delay profile with $\delta_d^2 = e^{-(d-1)} / \sum_{u=1}^{D} e^{-(u-1)}$ and $D = 10$ taps, where $d = 1, \ldots, D$. The results are shown in Figs. 4.8, 4.9, and 4.10, where the parameters are chosen as: $L = 2$, $m = 1$, and BPSK in Fig. 4.8; $L = 4$, $m = 1$, and QPSK in Fig. 4.9; $L = 8$, $m = 1$, and 8-PSK in Fig. 4.10a; $L = 16$, $m = 1$, and 16-PSK in Fig. 4.10b. To see how OFDM-IM compares with classical OFDM, the

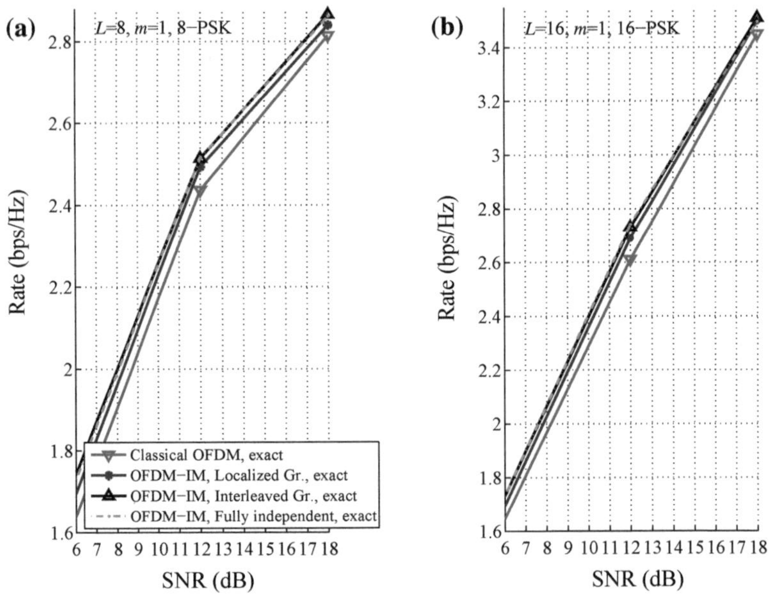

Fig. 4.10 Comparison between classical OFDM and OFDM-IM with localized grouping and interleaved grouping in terms of achievable rate with parameters chosen as: **a** $L = 8, m = 1$, 8-PSK; **b** $L = 16, m = 1$, 16-PSK

performance of classical OFDM employing the same alphabet for symbol modulation as OFDM-IM is also depicted. It is noted that the OFDM-IM with $L = 2$, $m = 1$, and localized grouping is exactly the enhanced subcarrier index modulation OFDM scheme [12]. Also note that the OFDM-IM and the classical OFDM in Figs. 4.8, 4.9, and 4.10 share the same entropy at the channel input. As seen from the figures, unlike classical OFDM for which the grouping type is irrelevant to performance, OFDM-IM with interleaved grouping can achieve up to 3, 2, 0.5, and 0.2 dB SNR gains over OFDM-IM with localized grouping at achievable rates equal to 0.95 bps/Hz in Fig. 4.8, 1.85 bps/Hz in Fig. 4.9, 2.6 bps/Hz in Fig. 4.10a, and 2.8 bps/Hz in Fig. 4.10b, respectively. This is expected because the independence of channel fading makes the different subcarrier activation types easier to differentiate, which in turn improves the coding gain from index modulation. On the other hand, at those achievable rates, OFDM-IM with interleaved grouping outperforms classical OFDM with an SNR gain of up to 3 dB. From the figures, we also see that the resulting achievable rates by interleaved grouping from the practical channel model match the ultimate ones derived by assuming fully independent fading, perfectly. This is in consistent with Proposition 3 as in those cases the interleaved grouping ensures the channel coefficients on subcarriers within a subcarrier group are approximately statistically independent.

4.2.2 Equiprobable Subcarrier Activation

In this subsection, we propose an ESA scheme, in which a table of a very small size is required to be stored at both the transmitter and the receiver.

Principle of ESA

Consider the OFDM-IM system introduced in Sect. 4.1.1. At the transmitter, a part of incoming bits, which is of number $p_1 = \lfloor \log_2(\mathrm{C}(L, \bar{m})) \rfloor$, are used to activate $\bar{m}(= L - m)$ out of L subcarriers per subcarrier group. For ease of decomposition, the mapping of p_1 bits to the subcarrier activation patterns (SAPs) for group g, i.e., $\Omega_i^g, i \in \{1, \ldots, 2^{p_1}\}$, is called unranking, and the reverse operation is called ranking. Without loss of generality, let us focus on the first subcarrier group and index the subcarriers belonging to it from $\{1, \ldots, L\}$. The superscript g is omitted for notational simplicity.

The existing solution, which is called combinatorial method, generates lexicographically ordered sequences [9]. Assume that the elements of Ω_i are sorted in a strictly increasing order. Then, according to the combinatorial theory we can build a one-to-one mapping which relates a natural number to $\bar{m}$-combinations:

$$i = \mathrm{C}(\Omega_i(\bar{m}) - 1, \bar{m}) + \cdots + \mathrm{C}(\Omega_i(1) - 1, 1) + 1. \tag{4.57}$$

To implement ranking, we first substitute Ω_i into (4.57) and then map the integer i to p_1 bits. For unranking, we first map the p_1 bits to an integer i and then determine $\Omega_i(\bar{m}), \ldots, \Omega_i(1)$ successively. Specifically, we choose $\Omega_i(\bar{m})$ to be the maximal integer that satisfies $i \geq \mathrm{C}(\Omega_i(\bar{m}) - 1, \bar{m}) + 1$ and $\Omega_i(\bar{m} - 1)$ to be the maximal integer satisfying $i \geq \mathrm{C}(\Omega_i(\bar{m}) - 1, \bar{m}) + \mathrm{C}(\Omega_i(\bar{m} - 1) - 1, 1) + 1$ and so on. An example for $L = 8$ and $\bar{m} = 2$ is given in Table 4.1.

The combinatorial method has some deficiencies. This can be observed from the example in Table 4.1 by examining the activation probabilities of all subcarriers, which are $\Pr(\mathrm{A}_1) = 3/8$, $\Pr(\mathrm{A}_2) = \Pr(\mathrm{A}_3) = \Pr(\mathrm{A}_4) = \Pr(\mathrm{A}_5) = \Pr(\mathrm{A}_6) = 5/16$, $\Pr(\mathrm{A}_7) = 1/16$, and $\Pr(\mathrm{A}_8) = 0$. This indicates that the lexicographically ordered subcarriers may be given totally different activation probabilities. However, since all subcarriers experience the same fading statistically, this mismatch will lead to a performance degradation to the system with ESA. In addition, another performance degradation is expected when we force $\log \Pr(\mathrm{A}_l) - \log \Pr(\bar{\mathrm{A}}_l) = \log(\bar{m}/m)$, $l = 1, \ldots, L$, for LLR detection in (4.9), which is often made in the literature, since it is violated in the practice. The above mentioned degradations motivate us to propose the ESA method.

The idea of ESA is to generate the SAPs in a cyclic manner. For better understanding, let us give an example. First, assume a legal SAP $\Omega_{b_1} = \{1, 2, \ldots, \bar{m}\}$. Then, by taking cyclic shift on each element of Ω_{b_1} based on $\{1, 2, \ldots, L\}$, which is called column cyclic shift in this letter, we can obtain $L - 1$ new patterns, which are $\{2, 3, \ldots, \bar{m} + 1\}, \{3, 4, \ldots, \bar{m} + 2\}, \ldots, \{L, 1, \ldots, \bar{m} - 1\}$. It is clear from the above L patterns that each subcarrier is activated with equal probability, namely $\bar{m}/L$.

Table 4.1 Combinatorial method versus ESA method ($L = 8, \bar{m} = 2$)

Incoming bits		Combinatorial method	ESA method
[0 0 0 0]	$\Omega_1 =$	$\{1, 2\}$	$\{1, 2\}$
[0 0 0 1]	$\Omega_2 =$	$\{1, 3\}$	$\{2, 3\}$
[0 0 1 0]	$\Omega_3 =$	$\{2, 3\}$	$\{3, 4\}$
[0 0 1 1]	$\Omega_4 =$	$\{1, 4\}$	$\{4, 5\}$
[0 1 0 0]	$\Omega_5 =$	$\{2, 4\}$	$\{5, 6\}$
[0 1 0 1]	$\Omega_6 =$	$\{3, 4\}$	$\{6, 7\}$
[0 1 1 0]	$\Omega_7 =$	$\{1, 5\}$	$\{7, 8\}$
[0 1 1 1]	$\Omega_8 =$	$\{2, 5\}$	$\{8, 1\}$
[1 0 0 0]	$\Omega_9 =$	$\{3, 5\}$	$\{1, 3\}$
[1 0 0 1]	$\Omega_{10} =$	$\{4, 5\}$	$\{2, 4\}$
[1 0 1 0]	$\Omega_{11} =$	$\{1, 6\}$	$\{3, 5\}$
[1 0 1 1]	$\Omega_{12} =$	$\{2, 6\}$	$\{4, 6\}$
[1 1 0 0]	$\Omega_{13} =$	$\{3, 6\}$	$\{5, 7\}$
[1 1 0 1]	$\Omega_{14} =$	$\{4, 6\}$	$\{6, 8\}$
[1 1 1 0]	$\Omega_{15} =$	$\{5, 6\}$	$\{7, 1\}$
[1 1 1 1]	$\Omega_{16} =$	$\{1, 7\}$	$\{8, 2\}$

Algorithm 2: ASDV generation

1: Initialization: $\mathbf{d}^{(1)} = [1, \ldots, 1, L - \bar{m} + 1]^T$, $I_1 = L, b = 1, j = \bar{m} - 1$
2: **while** $d^{(b)}(1) \leq L/\bar{m}$
3: $\quad b = b + 1$
4: $\quad d^{(b)}(1 : \bar{m} - 2) = d^{(b-1)}(1 : \bar{m} - 2)$
5: $\quad d^{(b)}(j) = d^{(b-1)}(j) + 1$
6: $\quad$ **for** $l = j + 1 : \bar{m} - 1$
7: $\qquad d^{(b)}(l) = 1$
8: $\quad$ **end for**
9: $\quad d^{(b)}(\bar{m}) = L - \sum_{j=1}^{\bar{m}-1} d^{(b)}(j)$
10: $\quad$ **if** $d^{(b)}(\bar{m}) < 1$, $j = j - 1$, Go to line 5 **end if**
11: $\quad$ Store $\mathbf{d}^{(b)}$ and calculate I_b if $\mathbf{d}^{(b)}$ is not a cyclic shift of $\{\mathbf{d}^{(1)}, \ldots, \mathbf{d}^{(b-1)}\}$.
12: **end while**
13: Output: $\mathbf{d}^{(1)}, \ldots, \mathbf{d}^{(\lceil T \rceil)}, I_1, \ldots, I_{\lceil T \rceil}$

Similarly, taking column cyclic shifts to $\Omega_{b_2} = \{1, 2, \ldots, \bar{m} - 1, \bar{m} + 1\}$ for L times allows us to obtain another N different patterns, which also satisfy the ESA requirement. This process continues until $2^{\lfloor \log_2 C(L,\bar{m}) \rfloor}$ different patterns are collected. It is worth noting that the above example only shows how to perform column cyclic shifts and ignores the problem of how to figure out the basic patterns such as Ω_{b_1} and Ω_{b_2}. The solution to this problem is given in the following. To begin with, let us define a so-called adjacent subcarrier distance vector (ASDV) as

---

Algorithm 3: Unranking

Initialization: Transfer p_1 bits into integer i, $b = 1$, $i' = i$
while $i' > I_b$
  $i' = i' - I_b$, $b = b + 1$
end while
$\Omega_1^{(i)} = i'$
for $j = 2$ to k
  $\Omega_i(j) = \Omega_i(j-1) + d^{(b)}(j-1)$
  if $\Omega_i(j) > L$
    $\Omega_i(j) = \Omega_i(j) - L$
  end if
end for
Output: Ω_i

$$\mathbf{d}^{(b)} = [d_1^{(b)}, \ldots, d_k^{(b)}]^T, \tag{4.58}$$

with $d_j^{(b)} =< \Omega_{<j+1>_{\bar{m}}}^{(b)} - \Omega_j^{(b)} >_L$, where $b \in \{1, \ldots, 2^{p_1}\}$ and $j \in \{1, \ldots, \bar{m}\}$. For instance, in the above example $\mathbf{d}^{(b_1)} = [1, \ldots, 1, L+1-\bar{m}]^T$, $\mathbf{d}^{(b_2)} = [1, \ldots, 1, 2, L-\bar{m}]^T$, and all column cyclic shifts generated by Ω_{b_1} and Ω_{b_2} share the same $\mathbf{d}^{(b_1)}$ and $\mathbf{d}^{(b_2)}$, respectively. By definition, we can now conclude that all ASDVs associated with the basic patterns have the same l_1 norm, which equals L. As a double check, in the above example $\left\|\mathbf{d}^{(b_1)}\right\|_1 = \left\|\mathbf{d}^{(b_2)}\right\|_1 = L$. Therefore, based on this property there will be $\mathrm{C}(L-1, \bar{m}-1)$ candidates for the basic patterns given L and $\bar{m}$. However, we should be careful since those candidates include a couple of ones that are cyclically shifted with each other, which will produce the same group of patterns. For example, for two ASDVs $[e, f, g]$ and $[g, e, f]$, where e, f, g are positive integers with $e + f + g = L$, their corresponding basic patterns are $\{1, 1+e, 1+e+f\}$ and $\{1, 1+g, 1+g+e\}$, respectively, where the latter is the g-times column cyclic shifted version of the former. Since there are at most $\bar{m}-1$ distinct cyclic shifts for a pattern, the number of valid candidates for the basic patterns approximates to $T = \mathrm{C}(L-1, \bar{m}-1)/\bar{m}$. From above, we summarize the generation of ASDVs in Algorithm 2. Since all ASDVs have the same l_1 norm of L, we only consider the case that the first element of any ASDV is no larger than $L/\bar{m}$, as reflected in line 2. With all ASDVs and the numbers of column cyclic shifts they can generate, denoted by $I_1, \ldots, I_{\lceil T \rceil}$, we can therefore proceed to perform ranking and unranking.

To determine the SAP according to the corrsponding p_1 bits, we first figure out which basic pattern it belongs to and then decide how many column cyclic shifts it should perform. This process is detailed in Algorithm 3, where the "while" loop determines the basic pattern, the variable i' stores the number of column cyclic shifts, and the "for" loop performs the column cyclic shifts.

Algorithm 4: Ranking

1: Initialization: $b' = 1$, $v = 0$
2: Calculate $\mathbf{d}^{(b)}$ from $\Omega_{\hat{i}}$ according to ASDV definition.
3: **while** $\mathbf{d}^{(b)} \neq \mathbf{d}^{(b')}$
4: $\quad v = v + I_{b'}, b' = b' + 1$
5: **end while**
6: $v = v + \Omega_{\hat{i}}(1)$
7: Output: The binary representation of v

Given a SAP $\Omega_{\hat{i}}$, the basic pattern it belongs to can be directly determined according to the definition of ASDV and the number of column cyclic shifts it results in can be also directly inferred from its first subcarrier index. The entire process of ranking is presented in Algorithm 4.

According to the above-mentioned principle, we also present the results of the ESA method for $L = 8$ and $\bar{m} = 2$ in Table 4.1. It is observed that all subcarriers are activated with an equal probability of $1/4$ for this example.

From above, we see that both the ranking and the unranking involve very simple mathematical operations and the computational complexity mainly arises from the generation of ASDVs. Fortunately, the ASDVs need to be prepared only once prior to the signal transmission and reception provided that they can be saved by both the transmitter and the receiver. Since the total ASDVs just include roughly $C(L-1, \bar{m}-1)$ indices, this storage requirement is easy to satisfy in today's systems. We should note that the proposed ESA method cannot always guarantee perfect equiproble subcarrier activation; however, it tries as best as it can. This imperfection occurs when 2^{p_1} is not a multiple of L. Fortunately, since in OFDM-IM systems the group size L is usually chosen to be an integer power of two, this imperfection happens very rarely.

SNR Gain Analysis

We focus on the g-th subcarrier group. However, for brevity we omit the superscript g for all variables. For notational simplicity, we let $\beta_l = l$ such that all subcarrier within the subcarrier group will be indexed from $\{1, \ldots, L\}$. Let $\Pr(\mathbf{x} \rightarrow \hat{\mathbf{x}})$ represent the PEP for the event that $\mathbf{x} = [x_1, \ldots, x_L]^T$ is transmitted; however, it is erroneously detected as $\hat{\mathbf{x}} = [\hat{x}_1, \ldots, \hat{x}_L]^T$. From [9], this PEP can be approximated as

$$\Pr\left(\mathbf{x} \rightarrow \hat{\mathbf{x}}\right) \cong \frac{1}{12} E_{\mathbf{h}}\left[e^{-q_1\|\Lambda\mathbf{h}\|_F^2}\right] + \frac{1}{4} E_{\mathbf{h}}\left[e^{-q_2\|\Lambda\mathbf{h}\|_F^2}\right], \tag{4.59}$$

where $q_1 = \gamma/4$, $q_2 = \gamma/3$, and $\Lambda = \mathrm{diag}\{\mathbf{x} - \hat{\mathbf{x}}\}$. Denote Ξ as an index set whose elements indicate the locations associated with the non-zero values of $\mathbf{x} - \hat{\mathbf{x}}$, $\tilde{\mathbf{h}}$ as the effective CFR vector whose entries are $\{h_i, i \in \Xi\}$, and $\tilde{\mathbf{X}}$ as a diagonal matrix of dimensions $|\Xi| \times |\Xi|$ whose diagonal elements are $\{\Lambda(i,i), i \in \Xi\}$. The expectations in (4.59) can be calculated as

$$E_{\mathbf{h}}\left[e^{-q_z\|\Lambda\mathbf{h}\|_F^2}\right] = \frac{1}{\pi^{|\Xi|}|\tilde{\Sigma}|}\int e^{-\tilde{\mathbf{h}}^H\mathbf{W}_z^{-1}\tilde{\mathbf{h}}}d\tilde{\mathbf{h}}$$
$$= |q_z\mathbf{A}\tilde{\Sigma} + \mathbf{I}_{|\Xi|}|^{-1},\ z = 1, 2, \tag{4.60}$$

where $\mathbf{A} = \tilde{\mathbf{X}}^H\tilde{\mathbf{X}}$, $\tilde{\Sigma} = E_{\tilde{\mathbf{h}}}\{\tilde{\mathbf{h}}\tilde{\mathbf{h}}^H\}$ is the channel covariance matrix, and $\mathbf{W}_z^{-1} = q_z\mathbf{A} + \tilde{\Sigma}^{-1}$. Finally, having the PEP and letting $N(\mathbf{x}, \hat{\mathbf{x}})$ measure the number of bits in error, we can obtain an upper bound on the ABEP of OFDM-IM systems as

$$P_e \le \frac{1}{p2^p}\sum_{\mathbf{x}}\sum_{\hat{\mathbf{x}}}\Pr(\mathbf{x} \to \hat{\mathbf{x}})\, N(\mathbf{x}, \hat{\mathbf{x}}), \tag{4.61}$$

where $p = p_1 + p_2$ with $p_2 = \bar{m}\log_2(M)$. Note that (4.61) applies to both the combinatorial and the ESA methods, leaving their difference in the selection of $\mathbf{x}$.

From (4.61), the asymptotic ABEP can be readily derived as

$$P_e \cong \frac{1}{p2^p}\sum_{\mathbf{x}}\sum_{\hat{\mathbf{x}}}\left(\frac{4^r}{12} + \frac{3^r}{4}\right)\prod_{\xi=1}^{r}\lambda_\xi^{-1}\gamma^{-r}N(\mathbf{x}, \hat{\mathbf{x}})$$
$$\cong \gamma^{-d_{min}}\frac{4^{d_{min}} + 3^{d_{min}+1}}{12p2^p}\sum_{d_{\min}}\prod_{\xi=1}^{d_{min}}\lambda_\xi^{-1}N(\mathbf{x}, \hat{\mathbf{x}}), \tag{4.62}$$

where $r = \text{rank}\{\mathbf{A}\tilde{\Sigma}\}$, $d_{min} = \min r$ is the diversity order, $\sum_{d_{min}}$ represents the sum operation applying to all $\mathbf{x}$ and $\hat{\mathbf{x}}$ conditioned on $r = d_{min}$, and λ_ξ, $\xi = 1, \ldots, r$, are the non-zero eigenvalues of $\mathbf{A}\tilde{\Sigma}$.

By definition, the coding gain can be derived from (4.62) as

$$G_c = \left(\frac{4^{d_{\min}} + 3^{d_{\min}+1}}{12p2^p}\sum_{d_{\min}}\prod_{\xi=1}^{d_{\min}}\lambda_\xi^{-1}e(\mathbf{x}, \widehat{\mathbf{x}})\right)^{-\frac{1}{d_{\min}}}, \tag{4.63}$$

which satisfies $P_e \cong (G_c\gamma)^{-d_{min}}$. From (4.63), the SNR gain of the ESA method over the combinatorial method can be readily expressed in terms of decibels as

$$G_{snr} = 10\log_{10}(G_c^{ESA}/G_c^{COM}), \tag{4.64}$$

where G_c^{ESA} and G_c^{COM} denote the resulting coding gains of the ESA and the combinatorial methods, respectively.

Performance Evaluation

We conduct computer simulations to examine the ABEP performance of the proposed ESA method. The system parameters are chosen as $N = 128$ and $L = 8$. The wireless channel is modeled by an uniform power delay profile with 10 taps. The localized grouping is assumed.

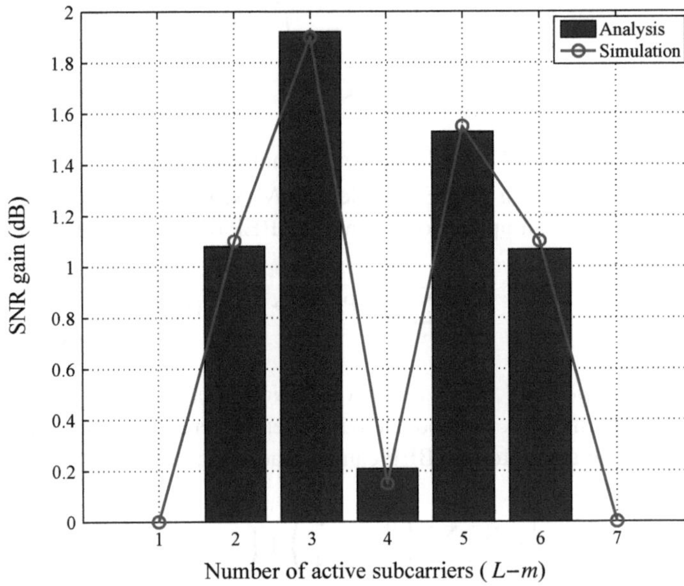

Fig. 4.11 The relationship between the SNR gain and the number of active subcarriers for $L = 8$

To completely isolate the effect of ESA, we first consider the OFDM-SSK system. By varying the number of active subcarriers $\bar{m}$, we obtain the SNR gains in Fig. 4.11, where the simulation results are obtained at a BER value of 10^{-5}. One can observe that an up to 1.9 dB SNR gain is approximately achieved for $\bar{m} = 3$, in which case the spectral efficiency is $R = 0.625$ bps/Hz. For $\bar{m} = 1$ and 7, no gain is attained since in both cases, all eight possible subcarrier activation patterns are used for indexing. Actually, since the ratio of the illegal subcarrier activation patterns is just about 8.6 %, a very small SNR gain, about 0.2 dB, is obtained for $\bar{m} = 4$. Finally, it should be noted that since the exact SNR gain depends on the specific probability distribution of the active subcarriers, asymmetric SNR gains are obtained.

We then consider the OFDM-IM system with BPSK, QPSK, 8-PSK, and 16-PSK modulations, which correspond to spectral efficiencies of $R = 1$, 1.375, 1.75, 2.125 bps/Hz, respectively, to examine the effect of ESA in the presence of modulated symbols. The comparison results for $\bar{m} = 3$ are shown in Fig. 4.12, where all analytical upper bounds are removed for clarity. The LLR detector exhibits nearly the same performance as the ML detector, and thus we only present an example for BPSK modulation in the figure. It can be seen from Fig. 4.12 that the SNR gains for the considered scenarios are approximately 1.3, 0.8, 0.6, and 0.3 dB, respectively. The decreasing SNR gain can be understood since the error contributed by modulation bits becomes dominant at high SNR and the ratio of the number of bits carried by the modulated symbols to that of the active subcarrier indices becomes higher as the modulation order increases. However, as the OFDM-IM scheme is beneficial for

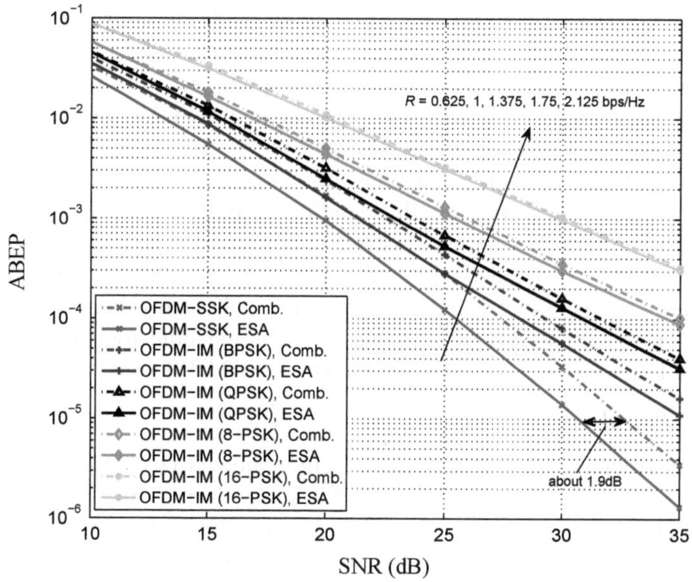

Fig. 4.12 Combinatorial method versus ESA method in terms of ABEP for OFDM-IM with BPSK modulation and OFDM-SSK with $L = 8$ and $\bar{m} = 3$

relatively low transmission rates below 2 bps/Hz [13], the ESA method is favorable in practice.

4.2.3 Hybrid Index Modulation

In this subsection, we propose hybrid subcarrier index modulation to further improve the spectral efficiency of OFDM-IM, in which different subcarrier groups can choose the OFDM-IM mode or OFDM mode, independently. Choice of the mode can be readily detected by the receiver and thus creating an additional means of information transfer.

Principle of the Proposed Scheme

With $L = M$, OFDM-IM achieves its maximum spectral efficiency, $\log_2(M)$, when only a single subcarrier is deactivated within each group. This is because $\mathrm{C}\,(L, m) \leq M^m$ with equality if and only if $m = 1$. In this case, OFDM-IM achieves a spectral efficiency which equals that of the classical OFDM system. However, with $m = 1$ and the same total transmit power, the OFDM-IM signal differs from the classical OFDM signal in two aspects: (i) OFDM-IM contains one less active subcarrier than classical OFDM; and (ii) the average power per active subcarrier is higher in the OFDM-IM system.

Consider now an OFDM-IM subcarrier group with the above-mentioned parameters and an OFDM subcarrier group with all subcarriers activated. One can then distinguish the two by determining whether the single inactive subcarrier is present or not. In other words, the "OFDM-IM" or "classical OFDM" mode of each group can be determined by checking the presence of the inactive subcarrier. Such "mode-selection" flexibility of each group introduces yet another means of information carriage: this mode can carry 1 bit, thus leading to spectral efficiency improvement.

This newly proposed system is illustrated in Fig. 4.13, where a group of $L = 4$ is exemplified. With $L = 4$, the modulation type is thereby QPSK according to $L = M$. The number of bits per block is increased from 8 to 9, boosting the spectral efficiency from 2 bps/Hz to 2.25 bps/Hz. The first bit is used as the indicator, which refers to the classical OFDM transmission if it is zero and the OFDM-IM transmission, otherwise.

Achievable Rate Analysis

The spectral efficiency of the proposed system is shown to be higher than that of the OFDM-IM system, but its superiority in the sense of error-free transmission rates is not that clear. To see how it is, we examine its achievable rate as follows. For ease of analysis, we assume Rayleigh fading channels and do not take into account the intercarrier interference (ICI). The ergodic achievable rate of the OFDM-HIM system can be readily derived in analogy with the analysis presented in Sect. 4.1.2, and thus is omitted here for brevity. In Fig. 4.14, the ergodic achievable rates of the OFDM-HIM system for BPSK and QPSK modulations, are compared with those of the classical OFDM and OFDM-IM systems. At very high SNR, all systems saturate

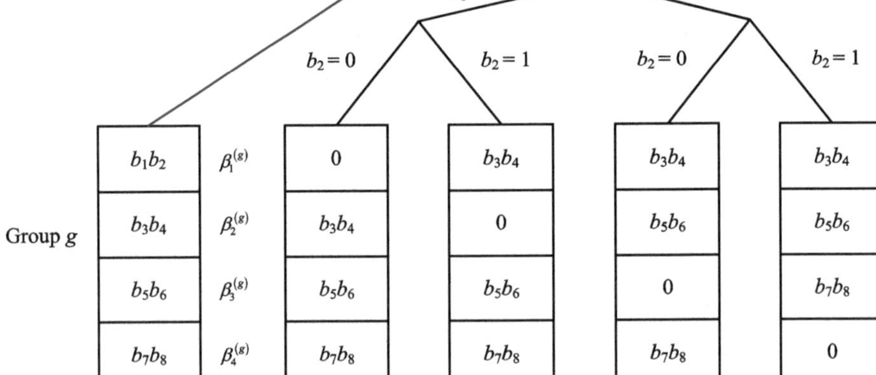

Fig. 4.13 Principle of the OFDM-HIM system

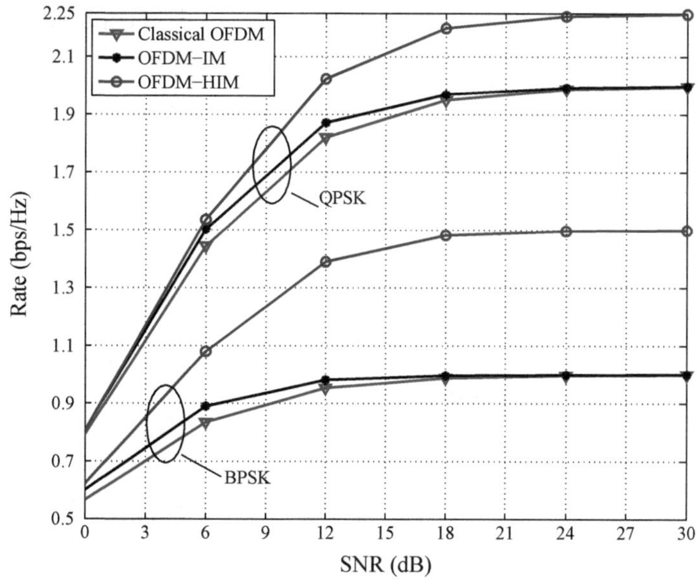

Fig. 4.14 Achievable rate of the OFDM-HIM system

at their corresponding uncoded spectral efficiencies as expected. However, in the low-to-medium SNR region, the superiority of the proposed system over the other two is still overwhelming. For example, the proposed system achieves an up to 3 dB SNR gain when the target rates are 0.9 and 1.9 bps/Hz for BPSK and QPSK modulations, respectively.

4.2.4 In-phase/Quadrature Index Modulation

In this subsection, we propose OFDM-I/Q-IM, which carries additional information bits through the in-phase and quadrature dimensions. The asymptotic ABEP and the exact coding gain achieved by OFDM-I/Q-IM are also analyzed.

Principle of the Proposed Scheme

The block diagram of the OFDM-I/Q-IM transmitter is depicted in Fig. 4.15. For the transmission of each OFDM block with N subcarriers, a total number of q incoming data bits are first divided into $2G$ groups, each of which consists of p bits, i.e., $q = 2pG$. Each group of p-bits is then used to generate the in-phase/quadrature part of an OFDM subblock consisting of $L = N/G$ subcarriers. Unlike classical OFDM that maps all data bits to constellation points for all subcarriers, each group of p-bits in OFDM-I/Q-IM is composed of two parts: the first part has p_1 bits used for index modulation and the second part has p_2 bits used for symbol modulation, i.e., $p = p_1 + p_2$. Specifically, for the in-phase/quadrature part of each OFDM subblock,

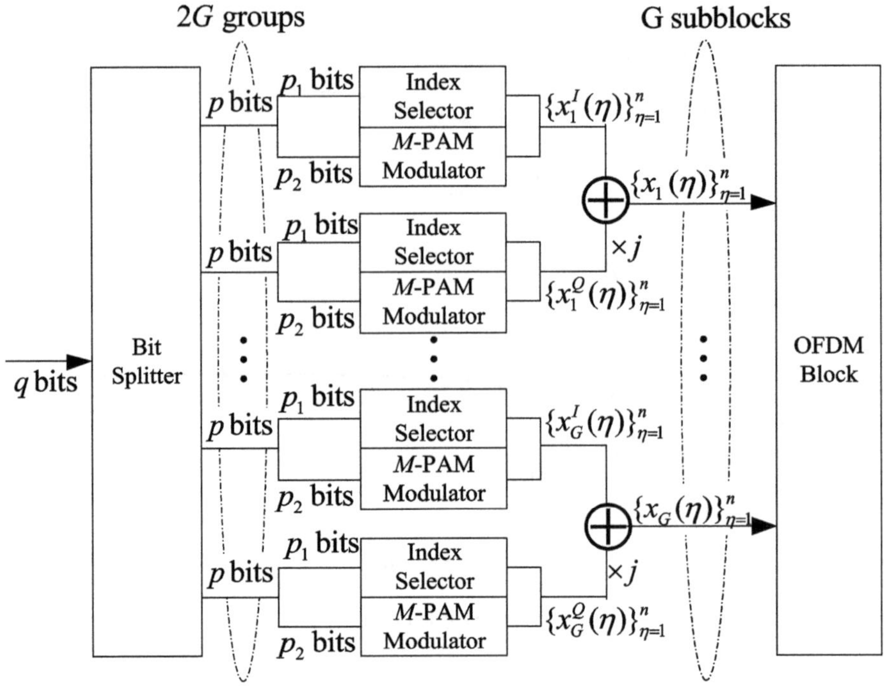

Fig. 4.15 Block diagram of OFDM-I/Q-IM transmitter

the p_1 bits are fed to the index selector to inactivate m out of L subcarriers, while the remaining $L - m$ subcarriers are set to be active. For the g-th ($1 \leq g \leq G$) OFDM subblock, the indices of $L - m$ active subcarriers in the in-phase and quadrature dimensions are independently given by

$$I_g^I = \left\{i_g^I(1), \ldots, i_g^I(L-m)\right\}, \quad I_g^Q = \left\{i_g^Q(1), \ldots, i_g^Q(L-m)\right\}, \tag{4.65}$$

where $i_g^I(a), i_g^Q(a) \in \{1, \ldots, L\}$ for $a = 1, \ldots, L-m$. The remaining p_2 bits are then fed to symbol modulator to generate transmitted symbols in the in-phase/quadrature dimension on the $L - m$ active subcarriers, resulting in the in-phase/quadrature part of the g-th OFDM subblock:

$$x_g^I(\eta) = \begin{cases} s_g^I(\eta), \ \eta \in I_g^I \\ 0, \ \text{otherwise} \end{cases}, \quad x_g^Q(\eta) = \begin{cases} s_g^Q(\eta), \ \eta \in I_g^Q \\ 0, \ \text{otherwise} \end{cases}, \tag{4.66}$$

where $s_g^I(\eta)$ and $s_g^Q(\eta)$ are independently drawn from an M-ary pulse amplitude modulation (PAM) alphabet Θ for $\eta = 1, \ldots, L$. Combining the in-phase and quadrature parts in (4.66), yields each complex OFDM subblock as

$$x_g(\eta) = x_g^I(\eta) + jx_g^Q(\eta),\ \ g = 1, \ldots, G,\ \eta = 1, \ldots, L. \tag{4.67}$$

Concatenating these G complex OFDM subblocks creates an OFDM block. Before transmission, the OFDM block is first transformed into the time domain signal block by utilizing an N-point inverse DFT, and then appended with a CP longer than the maximum delay spread of the channel.

After the removal of the CP at the receiver, an N-point DFT is employed to obtain the received block in the frequency domain, whose g-th subblock is given by

$$y_g(\eta) = h_g(\eta)x_g(\eta) + w_g(\eta),\ \ \eta = 1, \ldots, L, \tag{4.68}$$

where $h_g(\eta)$ and $w_g\,(\eta)$ denote the CFR and the zero-mean complex AWGN with variance σ^2 on the η-th tone of the g-th subblock, respectively. The spectral efficiency of OFDM-I/Q-IM is given by

$$F = \frac{2\,(p_1 + p_2)}{L} = 2\frac{\lfloor \log_2(\mathrm{C}(L, m)) \rfloor}{L} + 2\frac{(L - m)\log_2(M)}{L}, \tag{4.69}$$

where $p_1 = \lfloor \log_2(\mathrm{C}(L, m)) \rfloor$ and $p_2 = (L - m)\log_2(M)$.

For each subblock, by considering a joint detection for the indices of the active subcarriers and the modulated symbols carried on, the ML detector for OFDM-I/Q-IM is given by

$$\begin{aligned} \{\hat{x}_g(\eta)\}_{\eta=1}^{L} &= \arg\min_{\{x_g(\eta)\}_{\eta=1}^{L}} \sum_{\eta=1}^{L} |y_g(\eta) - h_g(\eta)x_g(\eta)|^2 \\ &= \arg\min_{\{x_g(\eta)\}_{\eta=1}^{L}} \sum_{\eta=1}^{L} |h_g(\eta)|^2 |r_g^I(\eta) - x_g^I(\eta)|^2 \\ &\quad + |h_g(\eta)|^2 |r_g^Q(\eta) - x_g^Q(\eta)|^2, \end{aligned} \tag{4.70}$$

where $r_g^Q(\eta)$ is the quadrature part of $r_g(\eta)$. From (4.70), it is clear that the detections of the in-phase and quadrature parts can be separated due to the mutual independence.

Performance Evaluation

The upper bounded and asymptotic ABEPs of OFDM-I/Q-IM systems can be derived in analogy with Sect. 4.2.3. For brevity, the analysis is omitted.

Figure 4.16 shows the comparison results of the ABEP performance for 4-QAM source. To make a fair comparison, all techniques operate at the same spectral efficiency of $F = 2$ bps/Hz, where $L = 2$ and $m = 1$ for the in-phase/quadrature part of OFDM-I/Q-IM and $L = 4$ and $m = 1$ for OFDM-IM. As seen from Fig. 4.16, OFDM-I/Q-IM achieves about 6 dB SNR gain over classical OFDM.

Figure 4.17 compares the coding gains of OFDM-I/Q-IM and OFDM-IM over classical OFDM, where three different spectral efficiency values are considered: 2, 4, and 6 bps/Hz. To achieve the same spectral efficiency as the classical OFDM with

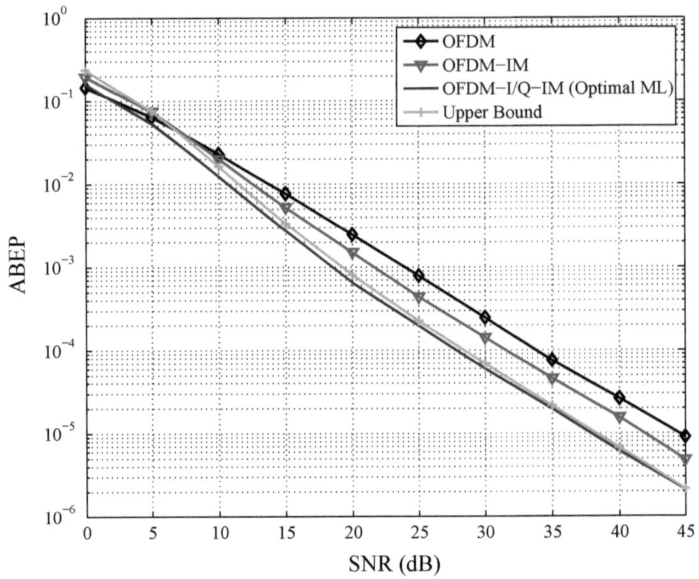

Fig. 4.16 ABEP performance of classical OFDM, OFDM-IM and OFDM-I/Q-IM at $\digamma = 2$ bps/Hz

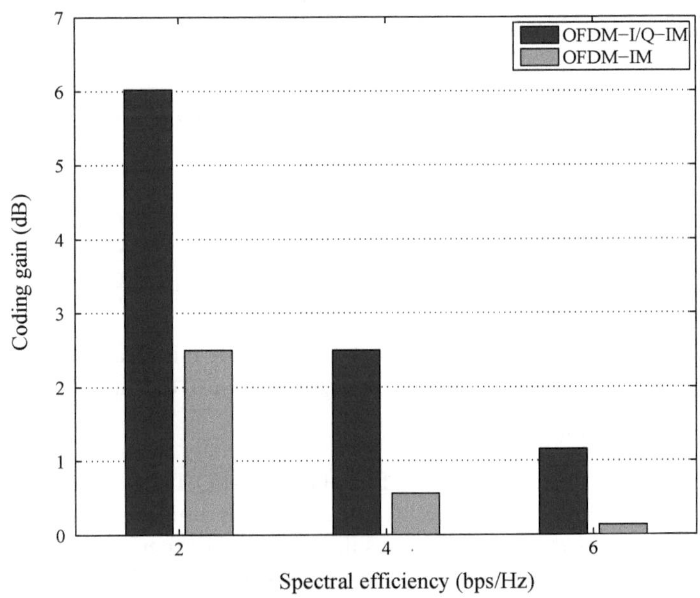

Fig. 4.17 Coding gains of OFDM-IM and OFDM-I/Q-IM over classical OFDM under different spectral efficiency values

M^2-QAM, OFDM-IM can simply deactivate one out of M^2 subcarriers deliberately. For OFDM-I/Q-IM, it can achieve the same spectral efficiency by deactivating one out of M subcarriers in the in-phase/quadrature dimension. It can be observed from Fig. 4.17 that both the coding gains of OFDM-I/Q-IM and OFDM-IM over classical OFDM decrease as the spectral efficiency value increases. However, at the same spectral efficiency, the coding gain achieved by OFDM-I/Q-IM is much larger than that achieved by OFDM-IM.

4.3 Applications to High-Mobility Communications

In this section, we propose a novel scheme which merges the advantages of the well-known ICI self-cancellation technique and the OFDM-IM technique to combat the ICI for V2X communications.

4.3.1 OFDM-IM with ICI Self-Cancellation

In this subsection, we first briefly review the ICI self-cancellation techniques and then present the idea of how to integrate ICI self-cancellation techniques into the OFDM-IM framework. The transceiver structure of the proposed system is designed, whose overall complexity is shown to be nearly the same as that of classical OFDM.

ICI Self-Cancellation Techniques Overview

ICI self-cancellation methods have been attracting considerable attention for its simplicity in implementation and effectiveness in ICI mitigation ever since its emergence in [14]. In this monograph, we refer to the ICI self-cancellation techniques as the ones that resort to data repetition in the frequency domain, though there are also other variations. Generally, ICI self-cancellation techniques are designed based on the single carrier frequency offset (CFO) model. Under the single CFO assumption, as indicated in Fig. 4.18a, b, the ICI coefficient from one subcarrier to the desired subcarrier, whose magnitude and phase refer to the attenuation and the rotation of the interference signal, is closely related to its adjacent subcarrier, symmetric subcarrier, and mirror subcarrier [15]. Such relationship is shown to be either nearly the same magnitude and opposite phases or nearly the same magnitude and a phase difference of π, depending on their frequency separation from the desired subcarrier as well as the CFO value. Based on this property, we can simply transmit data copies, either with sign conversion or phase conjugation relationship, onto the adjacent, symmetric, or mirror subcarrier pair of the OFDM symbol, and expect a reduced ICI level after combination at the receiver. This is the key idea of ICI self-cancellation.

Some specific properties of the ICI self-cancellation techniques are:

- The spectral efficiency is halved with respect to classical OFDM, yet can be compensated for by employing larger signal alphabet sizes.

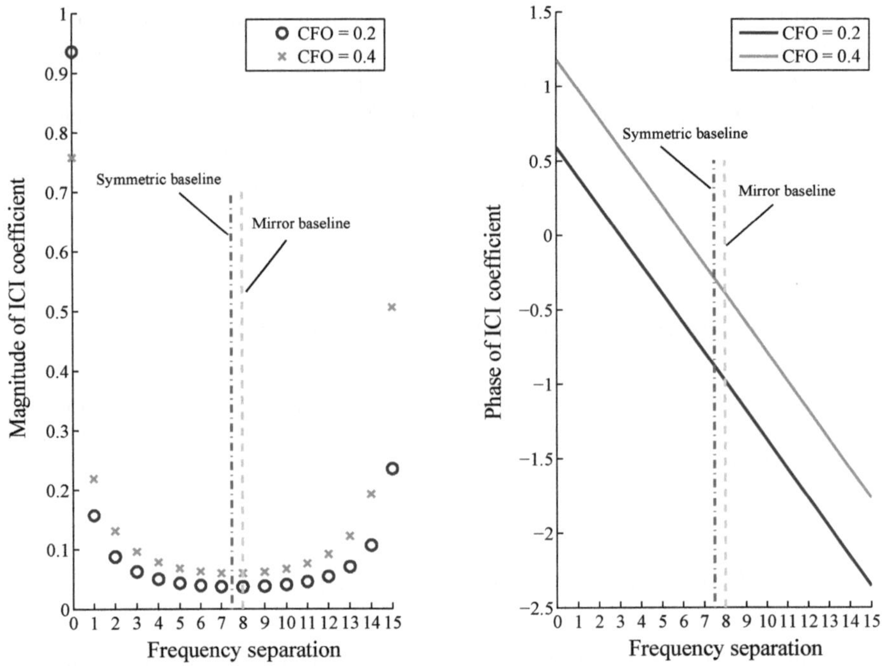

(a) Magnitude of ICI coefficient (b) Phase of ICI coefficient

Fig. 4.18 ICI self-cancellation techniques: **a** Plot of magnitudes of ICI coefficients for CFO equal to 0.2 and 0.4. **b** Plot of phases of ICI coefficients for CFO equal to 0.2 and 0.4

- With respect to the one-path approach, the two-path approach is compatible with the traditional OFDM transceiver design without any modifications, but is prone to CFO deviation.
- The performance of different schemes varies with different CFO values and channel conditions, and thus scheme selection in practice depends on the actual communication environment. For example, the carrier-to-interference ratio (CIR) performance of conversion-based schemes is degraded with a large channel length.

Integration of ICI Self-Cancellation into OFDM-IM

OFDM-IM transfers the part of information bits from the modulated symbols to their indices, which are transmitted implicitly over the superimposed signal generated from the remaining subcarriers, maintaining the spectral efficiency of the system and meanwhile having a smaller number of active subcarriers than classical OFDM at any time. This provides OFDM-IM with the potential of ICI reduction. However, in high-speed mobile channels such as V2X channels where ICI can be severe, the received signal power of each inactive subcarrier will be enhanced, whereas that of each active subcarrier will be reduced, rendering subcarrier states difficult to identify and further leading to worse system performance than classical OFDM. ICI self-cancellation techniques provide a possible solution to this problem since, as discussed above, they will reduce the power leakage from the active subcarriers to the inactive subcarriers due to ICI. In addition, the performance of ICI self-cancellation should be improved due to a smaller number of active subcarriers by the nature of OFDM-IM. Therefore, we expect a win-win situation as long as the ICI self-cancellation techniques can be successfully integrated into the OFDM-IM architecture. We consider the one-path implementation of the integration. For brevity, we take Zhao's scheme [14], which applies adjacent mapping and conversion operation, as an illustrative example.

Suppose that the system bandwidth is B_T, which is occupied by a total of N OFDM subcarriers. The subcarrier spacing is $B_S = B_T/N$. In the proposed scheme, the N subcarriers are split into $G = N/2L$ groups, each of which consists of $2L$ subcarriers. For subcarrier group g, the subcarrier indices are given by

$$\Psi^g = \left\{\beta_1^g, \beta_2^g, \ldots, \beta_{2L}^g\right\}, \tag{4.71}$$

where $\beta_l^g \in \{1, \ldots, N\}$ with $g = 1, \ldots, G$ and $l = 1, \ldots, 2L$. Note that $\Psi^{g_1} \cap \Psi^{g_2} = \emptyset$ and $\Psi^1 \cup \cdots \cup \Psi^G = \{1, \ldots, N\}$, where $g_1, g_2 \in 1, \ldots, G$ and $g_1 \neq g_2$. Here, we consider the case that the subcarriers of indices $\left\{\beta_1^g, \beta_3^g, \ldots, \beta_{2L-1}^g\right\}$ follow the interleaved mapping, i.e., $\beta_{2t-1}^g = 2(Lg - L + t) - 1$, and let $\beta_{2t}^g = \beta_{2t-1}^g + 1$, where $t = 1, \ldots, L$, in order to allow ICI to be self cancelled.

Figure 4.19 depicts the transmitter structure of the proposed scheme. First, a total of q bits are equally split into G blocks, each of which consists of two parts p_1 and p_2. This means $G(p_1 + p_2) = q$. Then, the first part p_1 will determine the states of all subcarriers and the second part p_2 will modulate the symbols transmitted on

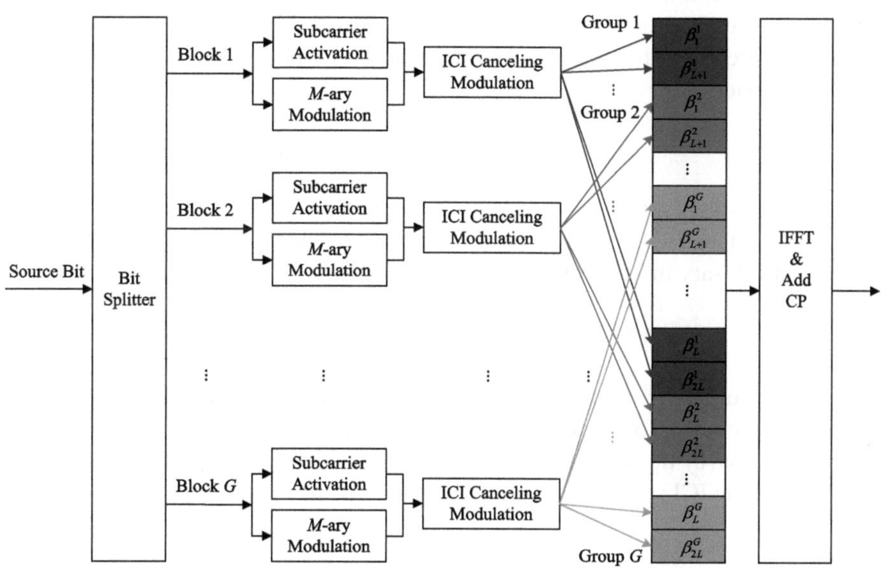

Fig. 4.19 Transmitter structure of the proposed scheme

the active subcarriers within a subcarrier group. Details on how they work will be presented in the following by taking block g as a demonstrative example, where $g \in 1, \ldots, G$.

The p_1 bits determine m out of L subcarriers of indices $\{\beta_1^g, \beta_3^g, \ldots, \beta_{2L-1}^g\}$ to be inactive and the other L subcarriers of indices $\{\beta_2^g, \beta_4^g, \ldots, \beta_{2L}^g\}$ follow the same states as those of indices $\{\beta_1^g, \beta_3^g, \ldots, \beta_{2L-1}^g\}$. Note that given L and m, there are in total $\mathrm{C}(L, m)$ combinations of active/inactive subcarrier indices. Take $L = 4$ and $m = 2$ for example. All $\mathrm{C}(4, 2) = 6$ combinations for subcarrier group g, i.e., $\{\beta_1^g, \beta_2^g, \ldots, \beta_{2L}^g\}$, are listed as follows:

$$
\begin{aligned}
\Omega_1^g &= \{\beta_1^g, \beta_2^g, \beta_3^g, \beta_4^g\} \left(\bar{\Omega}_1^g = \{\beta_5^g, \beta_6^g, \beta_7^g, \beta_8^g\}\right), \\
\Omega_2^g &= \{\beta_1^g, \beta_2^g, \beta_5^g, \beta_6^g\} \left(\bar{\Omega}_2^g = \{\beta_3^g, \beta_4^g, \beta_7^g, \beta_8^g\}\right), \\
\Omega_3^g &= \{\beta_1^g, \beta_2^g, \beta_7^g, \beta_8^g\} \left(\bar{\Omega}_3^g = \{\beta_3^g, \beta_4^g, \beta_5^g, \beta_6^g\}\right), \\
\Omega_4^g &= \{\beta_3^g, \beta_4^g, \beta_5^g, \beta_6^g\} \left(\bar{\Omega}_4^g = \{\beta_1^g, \beta_2^g, \beta_7^g, \beta_8^g\}\right), \\
\Omega_5^g &= \{\beta_3^g, \beta_4^g, \beta_7^g, \beta_8^g\} \left(\bar{\Omega}_5^g = \{\beta_1^g, \beta_2^g, \beta_5^g, \beta_6^g\}\right), \\
\Omega_6^g &= \{\beta_5^g, \beta_6^g, \beta_7^g, \beta_8^g\} \left(\bar{\Omega}_6^g = \{\beta_1^g, \beta_2^g, \beta_3^g, \beta_4^g\}\right)
\end{aligned}
$$

where $g = 1, \ldots, G$, Ω_j^g represents the j-th combination of the active subcarrier indices for subcarrier group g, and $\bar{\Omega}_j^g$ is the complement of Ω_j^g with respect to the set Ψ^g, which therefore denotes the j-th combination of the inactive subcarrier indices for subcarrier group g. Usually, $\mathrm{C}(L, m)$ is not a power of 2. For ease of implementation, only $2^{\lfloor \log_2(\mathrm{C}(L,m)) \rfloor}$ combinations are to be taken into account. Under this constraint, we have $p_1 = \lfloor \log_2(\mathrm{C}(L, m)) \rfloor$. The way to select one out of $2^{\lfloor \log_2(\mathrm{C}(L,m)) \rfloor}$ combinations according to random bits has been provided in Sect. 4.2.2 and it is omitted here for brevity. Assume that the output of the index selector for active subcarriers is

$$
\mathbf{x}_c^g \in \left\{\Omega_1^g, \Omega_2^g, \ldots, \Omega_{2^{\lfloor \log_2(\mathrm{C}(L,m)) \rfloor}}^g\right\}. \tag{4.72}
$$

The p_2 bits generate $(L - m)$ M-ary modulated symbols, which implies that $p_2 = (L - m) \log_2(M)$. Assume that the vector of the modulated symbols at the output of the M-ary modulator is

$$
\tilde{\mathbf{x}}_s^g = \left[s_1^g, s_2^g, \ldots, s_{L-m}^g\right]^T \in \mathcal{X}^{L-m}. \tag{4.73}
$$

Then, the outputs, $\mathbf{x}_c^g$ in (4.72) and $\tilde{\mathbf{x}}_s^g$ in (4.73), are fed into the ICI self-canceling modulation module. Consider Zhao's scheme [14]. That is, a pair of adjacent subcarriers will convey the same information symbol with different polarity. Therefore, the output of the ICI self-canceling modulation module is of the form

$$
\mathbf{x}_s^g = \left[s_1^g, -s_1^g, s_2^g, -s_2^g \ldots, s_{L-m}^g, -s_{L-m}^g\right]^T. \tag{4.74}
$$

The elements of $\mathbf{x}_s^g$ are the transmitted signals on the active subcarriers of indices $\mathbf{x}_c^g$.

Denote the total transmit power of the system as P_T. By taking into account $\mathbf{x}_c^g$ and $\mathbf{x}_s^g$ for all g and concatenating G subcarrier groups, the frequency-domain OFDM symbol is created as

$$\mathbf{x} = [x_1, x_2, \ldots, x_N]^T, \tag{4.75}$$

where $x_i \in \{0, \sqrt{P_G/2(L-m)}\mathbf{x}_s^1, \ldots, \sqrt{P_G/2(L-m)}\mathbf{x}_s^G\}$ with $P_G = P_T/G$ representing the total transmit power within each subcarrier group, and $i = 1, \ldots, N$. Note that as the transmitter has no knowledge of channel state information, it is wise to allocate active subcarriers with equal power. Also note that the power allocated to each active subcarrier is $P_G/2(L-m)$ for OFDM-IM rather than $P_G/2L$ for classical OFDM in order to balance the total transmit power.

Before transmission, the IFFT is applied to (4.75), yielding

$$\breve{x}_k = \frac{1}{\sqrt{N}} \sum_{i=1}^{N} x_i e^{\frac{j2\pi(i-1)(k-1)}{N}}, \quad k = 1, \ldots, N, \tag{4.76}$$

and a length-Q cyclic prefix (CP) of samples $\left[\breve{x}_{N-Q+1}, \breve{x}_{N-Q+2}, \ldots, \breve{x}_N\right]^T$ is appended to the beginning of the time-domain OFDM symbol $\breve{\mathbf{x}} = \left[\breve{x}_1, \breve{x}_2, \ldots, \breve{x}_N\right]^T$.

Assume that the signal passes through a time varying multipath fading channel, whose d-th tap at the n-th time instant is denoted by $\breve{h}_d(n)$, where $n = 1, \ldots, N$ and $d = 1, \ldots, D$ with D representing the number of channel taps, and the CP length Q is no smaller than $D-1$. Figure 4.20 shows the receiving procedure of the

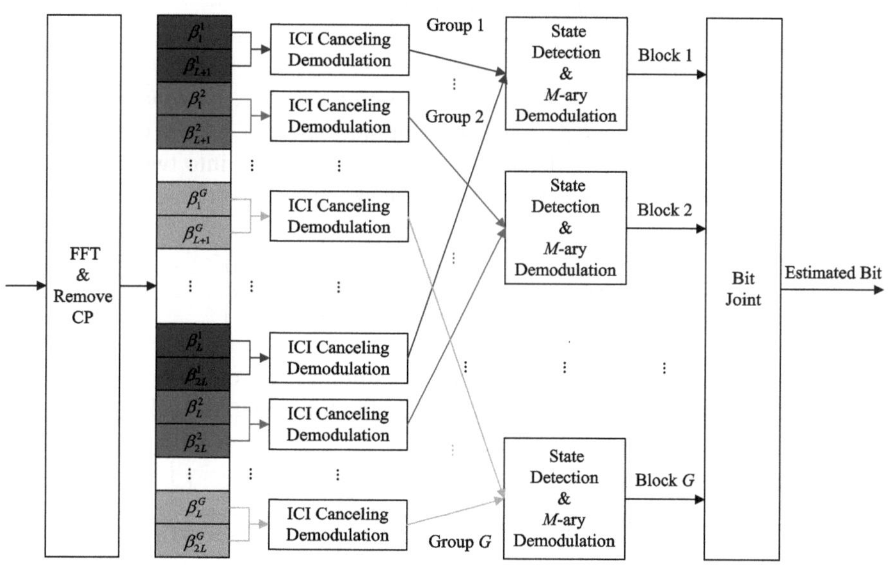

Fig. 4.20 Receiver structure of the proposed scheme

proposed scheme. At the receiver, the CP of the received signal is first removed and the application of the FFT is followed. The transformed frequency-domain received signal on the i-th ($i = 1, \ldots, N$) subcarrier is thereby given by

$$y_i = x_i h_i(0) + \underbrace{\sum_{k=1,k\neq i}^{N} x_i h_i(i-k)}_{\text{ICI}} + w_i, \tag{4.77}$$

where w_i is the AWGN of power σ^2 on the i-th subcarrier and $h_i(i-k) = \frac{1}{N}\sum_{d=1}^{D}\sum_{n=1}^{N}\breve{h}_d(n)\, e^{-\frac{j2\pi(i-k)(n-1)}{N}} e^{-\frac{\sqrt{-1}2\pi(k-1)(d-1)}{N}}$. Here, $\sigma^2 = N_0 B_S$ with N_0 denoting the power spectral density of the AWGN. In the remaining procedure, different operations from those in OFDM systems are performed. For brevity, we take subcarrier group g as a demonstrative example in the following, where $g \in 1, \ldots, G$.

First, the received signals on every two adjacent subcarriers are combined according to the ICI self-cancelling principle, yielding

$$\tilde{y}_{\beta_{2t-1}^g} = \frac{1}{2}\left(y_{\beta_{2t-1}^g} - y_{\beta_{2t}^g}\right), \ t = 1, \ldots, L \tag{4.78}$$

where we have simply neglected the fading diversity between two adjacent subcarriers. For the following decoding purpose, let $\tilde{y}_{\beta_{2t}^g} = 0$ with $t = 1, \ldots, L$.

Then, the joint detection of the active/inactive subcarrier indices and the modulated symbols is carried out based on the ML criterion:

$$\left(\hat{\mathbf{x}}_c^g, \hat{\mathbf{x}}_s^g\right) = \arg\min_{\mathbf{x}_c^g, \mathbf{x}_s^g} \sum_{i\in\Psi^g} \left|\tilde{y}_i - \tilde{h}_i x_i\right|^2 \tag{4.79}$$

where $\tilde{h}_{\beta_{2t-1}^g} = (\hat{h}_{\beta_{2t-1}^g} + \hat{h}_{\beta_{2t}^g})/2$ and $\tilde{h}_{\beta_{2t}^g} = 0$ with $t = 1, \ldots, L$ and $\hat{h}_i$ representing the channel estimate on the i-th subcarrier. Similar to Sect. 4.1.1, we can reduce the computational complexity by decomposing the ML detector into two steps. In the first step, we collect candidates for the modulated symbols as follows:

$$\hat{s}_t^g = \arg\min_{s\in\mathcal{S}} \left|\tilde{y}_{\beta_{2t-1}^g} - \tilde{h}_{\beta_{2t-1}^g}\sqrt{\frac{P_G}{2(L-m)}}s\right|^2. \tag{4.80}$$

Note that the outputs of (4.80) are L symbol estimates, but only $(L-m)$ ones are necessary. In the second step, we first calculate the following ML values:

$$\eta_t^g = 2\Re\left\{\tilde{y}_{\beta_{2t-1}^g}^* \tilde{h}_{\beta_{2t-1}^g}\sqrt{\frac{P_G}{2(L-m)}}\hat{s}_t\right\} - \frac{P_G\left|\tilde{h}_{\beta_{2t-1}^g}\hat{s}_t\right|^2}{2(L-m)}. \tag{4.81}$$

Then, we decide on one hand on the $(L - m)$ active indices who have maximum ML values in (4.81), obtaining $\hat{\mathbf{x}}_c^g$, and on the other hand on the $(L - m)$ symbol estimates in (4.80) with indices belonging to $\hat{\mathbf{x}}_c^g$, obtaining $\hat{\mathbf{x}}_s^g$. It is clear from (4.80) and (4.81) that the computational complexity of the two-step ML detector is about $\mathcal{O}(M)$ per subcarrier, which is the same as that of the classical OFDM detector.

Finally, the receiver de-maps $\left(\hat{\mathbf{x}}_c^g, \hat{\mathbf{x}}_s^g\right)$ to the corresponding information bits in reverse operations to the index mapper and M-ary modulation for subcarrier group g, and then collects all information bits from G subcarrier groups to complete reception of an OFDM symbol.

4.3.2 Performance Evaluation

In this subsection, we evaluate the performance of the proposed scheme in V2X channels.

So far, V2X channels have been extensively investigated. In the simulations, we adopt the channel model proposed in [16], which typically regarded as a standard V2X channel model dedicated for IEEE 802.11p. The measurement campaign was carried out in the metropolitan Atlanta, Georgia area including six scenarios, which are

- Scenario 1: V2V Expressway Oncoming;
- Scenario 2: V2V Expressway Same Direction with Wall;
- Scenario 3: V2V Urban Canyon Oncoming;
- Scenario 4: Roadside-to-vehicle (R2V) Expressway;
- Scenario 5: R2V Urban Canyon Oncoming; and
- Scenario 6: R2V Suburban Street.

The channel parameters for these six scenarios are listed in Table 4.2. For brevity, we only present the results under Scenario 1, Scenario 3, and Scenario 5, which cover both vehicle to vehicle (V2V) and road to vehicle (R2V) channels and a wide range of Doppler shifts.

Table 4.2 V2X channel parameters

Scenario	Velocity (km/h)	Doppler shift (Hz)
Scenario 1	104	1000–1200
Scenario 2	32–48	300
Scenario 3	104	900–1150
Scenario 4	104	600–700
Scenario 5	32–48	400–500
Scenario 6	32–48	300–500

In the simulations, the channel estimation is performed by the preamble, which is located prior to each OFDM data symbol and comprises an entire OFDM symbol known to both the transmitter and the receiver. Then, the channel estimates are used for the equalization and demodulation of the OFDM data symbol. For the proposed scheme, the system parameters are chosen as: (1) $G = 8$, $L = 4$, $m = 3$, and QPSK modulation; (2) $G = 4$, $L = 8$, $m = 6$ and QPSK modulation. For comparison, we choose

- Scheme 1: OFDM with BPSK modulation;
- Scheme 2: OFDM with ICI self-cancellation and QPSK modulation;
- Scheme 3: OFDM-IM with four subcarriers as a group, two subcarriers inactivated, and BPSK modulation.

Note that all schemes achieve the same spectral efficiency, i.e., 1 bps/Hz (without taking into account the null subcarriers and pilots), which means the comparison is fair in this sense.

Figure 4.21 shows the performance comparison results between Schemes 1–3 and the proposed scheme ($G = 8$, $L = 4$, $m = 3$, and QPSK modulation) under different scenarios. As can be seen, when SNR is low, Schemes 1 and 2 perform better than the OFDM-IM and the proposed scheme. This is because when the SNR is low, the noise largely contributes to the power of received signals, such that the power of the received signal at each inactive subcarrier becomes comparable to that at each active subcarrier, misleading the detection of subcarrier active statues. To make the

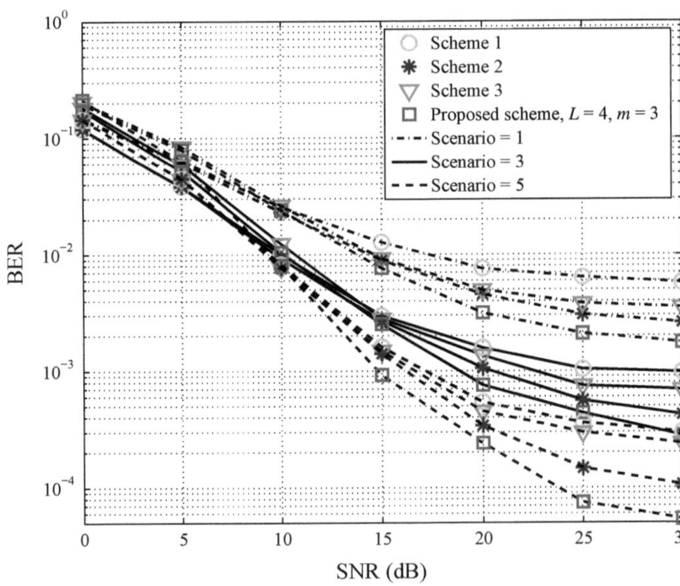

Fig. 4.21 BER performance comparison for Scenario 1, Scenario 3, and Scenario 5 among Schemes 1–3 and the proposed scheme ($L = 4$, $m = 3$, and QPSK modulation)

proposed scheme more practical, when SNR is below 10 dB, all subcarriers pairs in the proposed scheme can be set to be active. In this manner, we avoid the detection of the subcarrier active statues at the receiver and the proposed scheme is equivalent to the OFDM with the ICI self-cancellation scheme.

As the SNR increases, it can be seen that OFDM-IM and the proposed scheme outperform the classical OFDM scheme. This happens because the subcarriers in OFDM-IM and the proposed scheme are grouped in an interleaved manner, which reduces the effect of correlated frequency-selective fading in V2X channels significantly. However, OFDM-IM performs worse than Scheme 2 and exhibits an error floor at 25 dB. It can be understood since the ICI caused by Doppler shift largely deteriorates the system performance of OFDM-IM, while it is suppressed by the ICI self-cancellation method in Scheme 2. As can be seen in Fig. 4.21, when BER falls below 10^{-2}, the proposed scheme performs better than classical OFDM and OFDM-IM under different scenarios, and more importantly, even better than OFDM with ICI self-cancellation. This is as expected because in the proposed scheme, with the ICI self-cancellation technique being integrated into the OFDM-IM frame work, almost $N(L-m)/L$ subcarriers transmit zero energy and the number of active subcarriers which actually incur ICI is reduced to $(m/L)N$. Note that the above-mentioned SNR threshold, i.e., 10 dB, is meaningful since at it the BER value is around 10^{-2}, which is required by most practical systems.

As detailed in Table 4.3, Scenario 1 has the highest Doppler shift. Correspondingly, the BER error floors are the highest for all three schemes. However, it is worth mentioning that the proposed scheme performs best when the Doppler shift is relatively high. This is because as the Doppler-induced ICI becomes dominant, the effect of ICI suppression in the proposed scheme becomes more obvious with the adoption of the ICI self-cancellation technique.

In proposed scheme, the change of symbol modulation size can be offset by the subcarrier activation parameters to maintain the same transmitted information rate. Therefore, alternative to the parameter set used to generate Fig. 4.21, one can use $L=8$, $m=6$, and QPSK modulation without affecting the transmitted information rate. The corresponding results are shown in Fig. 4.22. From Fig. 4.22, we observe similar phenomenon to Fig. 4.21.

Table 4.3 System parameters of V2X channels

Parameter	Value
Signal bandwidth	10 MHz
Carrier frequency	5.9 GHz
Number of total subcarriers	64
Number of cyclic prefix	16
Subcarrier spacing	156.3 kHz
OFDM symbol duration	8 μs

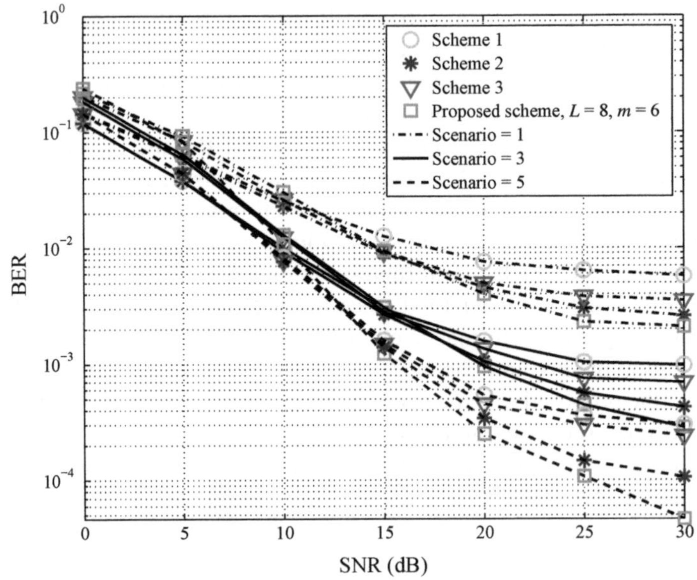

Fig. 4.22 BER performance comparison for Scenario 1, Scenario 3, and Scenario 5 among Schemes 1–3 and the proposed scheme ($L = 8$, $m = 6$ and QPSK modulation)

4.4 Summary

In this chapter, we introduced a novel frequency-domain index modulation scheme, called OFDM-IM. The achievable rate of OFDM-IM was analyzed, which provides insights into its strong potentials. To further improve the diversity gain, the coding gain, and the spectral efficiency of OFDM-IM, the interleaved grouping approach, the ESA method, the OFDM-HIM scheme, and the OFDM-I/Q-IM scheme were proposed. Finally, the idea of integrating ICI self-cancellation techniques into the OFDM-IM framework was clarified, and the superior ICI mitigation performance of the proposed scheme was verified in V2X channels.

References

1. D. Tse, P. Viswanath, *Fundamentals of Wireless Communications* (Cambridge University Press, Cambridge, 2005)
2. H. Bolcskei, D. Gesbert, A.J. Paulraj, On the capacity of OFDM-based spatial multiplexing systems. IEEE Trans. Commun. **50**(2), 225–234 (2002)
3. T.M. Cover, J.A. Thomas, *Elements of Information Theory*, 2nd edn. (Wiley, New York, 2006)
4. W. Zeng, C. Xiao, M. Wang, J. Lu, Linear precoding for finite-alphabet inputs over MIMO fading channels with statistical CSI. IEEE Trans. Signal Process. **60**(6), 3134–3148 (2012)

5. I.S. Gradshteyn, I.M. Rhyhik, *Table of Integrals, Series and Products*, 7th edn. (Academic, New York, 2007)
6. J.M. Kahn, J.R. Barry, Wireless infrared communications. Proc. IEEE **85**(2), 265–298 (1997)
7. Y. Yang, B. Jiao, Information-guided channel-hopping for high data rate wireless communication. IEEE Commun. Lett. **12**(4), 225–227 (2008)
8. M. Di Renzo, H. Haas, Bit error probability of SM-MIMO over generalized fading channels. IEEE Trans. Veh. Technol. **61**(3), 1124–1144 (2012)
9. E. Basar, U. Aygolu, E. Panayirci, H.V. Poor, Orthogonal frequency division multiplexing with index modulation. IEEE Trans. Signal Process. **61**(22), 5536–5549 (2013)
10. J. Harshan, B.S. Rajan, On two-user Gaussian multiple access channels with finite input constellations. IEEE Trans. Inf. Theory **57**(3), 1299–1327 (2011)
11. Telesystem Innovations, LTE in a Nutshell: System Overview. White Paper (2010)
12. D. Tsonev, S. Sinanovic, H. Haas, Enhanced subcarrier index modulation (SIM) OFDM, in *Proceedings of the IEEE Global Communications Conference (GLOBECOM) Workshops, Houston, TX, USA* (2011), pp. 728–732
13. N. Ishikawa, S. Sugiura, L. Hanzo, Subcarrier-index modulation aided OFDM - will it work? IEEE Access **4**, 2580–2593 (2016)
14. Y. Zhao, S.G. Häggman, Intercarrier interference self-cancellation scheme for OFDM mobile communication systems. IEEE Trans. Commun. **49**(7), 1185–1191 (2001)
15. M. Wen, X. Cheng, X. Wei, B. Ai, B. Jiao, A novel effective ICI self-cancellation method, in *Proceedings of the IEEE GLOBECOM, Houston, TX, USA* (2011), pp. 1–5
16. G. Acosta-Marum, M.A. Ingram, Six time- and frequency-selective empirical channel models for vehicular wireless LANs. IEEE Veh. Technol. Mag. **2**(4), 4–11 (2007)

Chapter 5
Conclusions and Future Directions

5.1 Conclusions

In this monograph, we investigated the space, space-time, and frequency domain index modulation techniques from three aspects, namely transceiver design, performance analysis, and performance optimization.

Firstly, we explored the potential of index modulation in the space domain. By extending the idea of SM to the receiver side, we proposed the GPQSM scheme, which allows activation of multiple receive antennas and further exploits both in-phase and quadrature dimensions. GPQSM generalizes the ZF/MMSE pre-coding aided SM schemes and conventional MIMO. The ABEP performance of GPQSM was analyzed and the superiority of GPQSM over conventional MIMO in terms of ABEP was revealed. The deficiency of ZF/MMSE pre-coding aided SM schemes lies in the resulting correlated channels and the constraint on the antenna configuration, which motivated us to propose the VSM scheme. Unlike all reported pre-coding aided SM schemes that exploit practical channels to convey information, VSM performs index modulation on virtual parallel channels resulting from the SVD of MIMO channels. The SVD relaxes the restriction on the antenna configuration, allowing VSM to be used as either a transmitter-side or a receiver-side SM scheme. To investigate the performance of VSM, tight upper bounded and asymptotic ABEPs were derived. Owing to less channel correlation, VSM significantly outperforms the existing precoding-aided SM schemes especially for relatively low transmission rates. To explore the potential of space domain index modulation techniques in the application to cooperative communication scenarios, we chose SSK modulation as a representative and discussed how it can be used in two-way AF relaying. A pragmatic communication strategy was proposed, in which the transceiver sends the signal by employing SSK modulation and detects the signal based on the MRC principle. By considering Nakagami-m fading channels, upper bounded and asymptotic ABEPs were derived. Moreover, the power allocation issue was also solved.

Secondly, we explored the potential of index modulation in the space-time domain. The DSM scheme, which exploits the antenna activation order to convey additional

M. Wen et al., *Index Modulation for 5G Wireless Communications*,
Wireless Networks, DOI 10.1007/978-3-319-51355-3_5

information, was proposed as a representative. DSM transmits space-time matrices that contain only one non-zero element in each row and each column, enabling differential modulation and demodulation. The ABEP performance of DSM was analyzed and a less-than-3dB performance loss was observed as compared with coherently detected SM. To solve the problem of high demodulation complexity, we proposed a low-complexity sub-optimal detector for DSM. The idea was motivated by the special structure of the information matrix, to which a Viterbi-like algorithm applies. Noticing that errors mainly arise from the differentiation between any two antenna activation orders having only two indices difference, we proposed the idea of Gray coding to minimize the performance loss. With Gray coding, any two antenna activation orders having only two indices difference always differ from only one bit, improving the coding gain achieved by the DSM system. Based on the Gray coding framework, we further proposed the I-gray code order to improve the diversity gain achieved by the DSM system, where the permutations of active antenna indices are selected from the odd (or even) positions of the full permutations in the gray code order. Apart from the transmitter-side DSM scheme, we further proposed the receiver-side DSM scheme, namely PDSM, which resorts to the ZF/MMSE precoding at the transmitter to activate a single receive antenna at each time instant and utilizes the activation order of the receive antennas for index modulation. The ABEP performance of PDSM was analyzed and a no more than 3dB performance loss was resulted as compared with PSM. To explore the potential of space-time domain index modulation techniques in the application to cooperative communication scenarios, we chose DSM and PDSM as representatives and discussed how they can be used in dual-hop AF relaying. We formed the DH-HDSM architecture by applying PDSM for the first hop and DSM for the second hop. To achieve high energy efficiency, single RN activation was ensured by either CD or DD protocols at the RNs, and in particular, a low-complexity near-optimal algorithm was proposed for DD. To evaluate the performance, ABEP upper bounds were provided for both CD and DD algorithms and the coding gain and diversity order of CD were analyzed. Based on the flexibility of RN configurations, we also analyzed the optimal systems setups to achieve either the optimum ABEP or the maximum energy efficiency.

Finally, we explored the potential of index modulation in the frequency domain. The OFDM-IM scheme, which conveys additional information through activation of subcarriers, was considered for this purpose. We for the first time conducted coded performance analysis for OFDM-IM, which takes into account both finite and infinite constellations input. It was revealed that OFDM-IM significantly outperforms plain OFDM in the case of finite constellation input especially for relatively low transmission rates and always performed worse in the case of infinite constellation input due to a loss of degrees of freedom. A useful lower bound on the achievable rate in closed form was also provided, which can be used to design the optimal subcarrier activation strategy, and it was found that the superiority of OFDM-IM over plain OFDM is greater for PSK than for QAM. To improve the diversity gain achieved by OFDM-IM, the interleaved grouping was proposed and shown to outperform the existing localized grouping with an SNR gain of up to 3dB. To improve the coding gain achieved by OFDM-IM, we proposed the ESA method, which allows equiprobable activation of subcarriers. The ESA method is preferred over the existing subcarrier activation methods especially for relatively low transmission rates, which happen to be favorable for OFDM-IM. The presence of empty subcarriers restricts the spectral efficiency of OFDM-IM systems. To solve this problem, we proposed the OFDM-HIM and OFDM-I/Q-IM schemes. OFDM-HIM introduces an additional bit for mode selection between OFDM-IM and plain OFDM in each subcarrier group, which can be detected by checking the presence of the inactive subcarrier. The achievable rate analysis for OFDM-HIM systems verified its advantages. OFDM-I/Q-IM further exploits the in-phase and quadrature dimensions for index modulation. It was verified that OFDM-I/Q-IM can harvest more than 6dB SNR gain over plain OFDM-IM under the same spectral efficiency. Since OFDM-IM is a variant of OFDM, it is susceptible to ICI. The key feature of OFDM-IM that partial subcarriers are kept inactive provides possibility to mitigate the ICI. To explore this potential, we integrated ICI self-cancellation techniques into the OFDM-IM framework and designed the variable OFDM-IM transceiver. Through simulations over V2X channels, the superior ICI mitigation performance of the proposed scheme was verified.

5.2 Future Directions

In this monograph, we mainly studied several representative index modulation techniques that exploit either the space, space-time, or frequency resources. However, there are still open problems.

Firstly, only Rayleigh fading channels are considered for the performance analysis of the PDQSM and VSM schemes. However, the practical channel may fit the Nakagami-m channel model for its generality. For this model, our analytical techniques face a great challenge since the joint PDF of the gains of the resulting parallel channels from pre-coding is difficult to characterize.

Secondly, the DSM scheme works for PSK input only. For QAM input, the current DSM design fails to guarantee the closure property under multiplication and requires increasing transmit power as more and more DSM blocks are transmitted.

Thirdly, the diversity order of the OFDM-IM system is limited to unity. The CIOD method is designed in the literature for plain OFDM-IM systems to improve the diversity order to two, but it cannot be directly applied to OFDM-I/Q-IM systems since the symbols carried on the real and imaginary parts of the subcarriers are drawn from independent PAM constellations.

On the other hand, there is also much research space for the development of new index modulation techniques.

Index modulation in the space domain can be also realized by grouping transmit antennas. This applies to the case that the number of transmit antennas is larger than that of receive antennas and should be coupled with pre-coding. With pre-coding, the number of resulting parallel channels is always equal to that of receive antennas as long as a group of transmit antennas whose number is no less than that of receive antennas are involved. Since different groups of transmit antennas involved for pre-coding result in different parallel channels, additional information bits can be modulated into the indices of groups, thus enhancing the spectral efficiency. However, the advantage is obtained at the sacrifice of diversity gain and this scheme may suffer from considerable ambiguity if different groups overlap significantly. To tradeoff the error performance and spectral efficiency, sophisticated grouping for the transmit antennas is needed. In addition, low-complexity near-optimal detection is demanding for this design.

Index modulation in the frequency domain can be also realized by assigning multiple modes to OFDM subcarriers. In current OFDM-IM systems, only two modes are available, which are null and conventional M-ary constellation, and the null mode itself cannot carry any information. Assigning more modes to a subcarrier is an effective means to improve the spectral efficiency. These modes should be as diverse as possible to ensure satisfactory error performance, which can be designed, for example, based on the minimum Euclidean distance criterion. The index modulation can be then performed by selecting the specific transmit modes associated with different subcarriers. Another realization of index modulation in the frequency domain can be the joint exploration of the in-phase and quadrature dimensions. This should not be mistaken with OFDM-I/Q-IM that explores the two dimensions independently. Indeed, indexing the elements across in-phase and quadrature dimensions allows us to modulate more bits. However, the error performance may be degraded without careful choosing the elements since the real and imaginary parts of the received signals are correlated. A wise method to circumvent this problem should discard those elements whose in-phase and quadrature components lie at the same subcarrier frequency.

Index modulation in the time domain is especially suitable for single-carrier transmission and fast fading channels. Recall that the single-carrier frequency domain equalization (SC-FDE) transmitted signal is related to the frequency-domain finite constellation input through IFFT. Discarding some samples of the SC-FDE signal will not affect the system performance much as the distortion can be compensated to some extent. The indices of discarded samples can be then used to carry additional information. On the other hand, the fast fading channel supports diversity gain for this index modulation system. For this method, the number of discarded samples should be carefully determined to guarantee an endurable distortion level; otherwise, the error performance may be seriously degraded since the distortion is not recoverable.

Index modulation in the space-frequency domain can be also applied to the receiver side via pre-coding. Since with pre-coding the received signals at different receive antennas are free of interchannel interference, the joint selection of the active elements in space and frequency domains will not increase the detection complexity as at the transmitter side. The detection at the receiver side is exactly the same as in OFDM-IM. In order to optimize the performance, we should minimize the correlation between the channel coefficients among the space-frequency blocks. When interleaved grouping is applied, which means the correlation among subcarriers in the same space-frequency block is minimized, the performance solely depends on the correlation among the antenna elements at the same subcarrier frequency. Since the pre-coding operation introduces correlation among channel coefficients at the same subcarrier frequency, the optimal performance can be obtained when the elements at the same subcarrier frequency are distributed to different groups.

MIX
Papier aus verantwortungsvollen Quellen
Paper from responsible sources
FSC® C105338

If you have any concerns about our products,
you can contact us on
ProductSafety@springernature.com

In case Publisher is established outside the EU,
the EU authorized representative is:
Springer Nature Customer Service Center GmbH
Europaplatz 3, 69115 Heidelberg, Germany

Printed by Libri Plureos GmbH
in Hamburg, Germany